Maria Auxiliadora Silva Oliveira
F. Hérica Chaves A.
Karlynne F. Mendonça

Prospecting botanical families of interest

Maria Auxiliadora Silva Oliveira
F. Hérica Chaves A.
Karlynne F. Mendonça

Prospecting botanical families of interest

The use of the Myrtaceae and Lamiaceae families in their antimicrobial potential

ScienciaScripts

Imprint

Cover image: www.ingimage.com

This book is a translation from the original published under ISBN 978-3-330-76952-6.

Publisher:
Sciencia Scripts
is a trademark of
Dodo Books Indian Ocean Ltd. and OmniScriptum S.R.L publishing group

120 High Road, East Finchley, London, N2 9ED, United Kingdom
Str. Armeneasca 28/1, office 1, Chisinau MD-2012, Republic of Moldova, Europe
Managing Directors: Ieva Konstantinova, Victoria Ursu
info@omniscriptum.com

Printed at: see last page
ISBN: 978-620-8-62413-2

CONTENTS

CHAPTER 1

THEORETICAL APPROACH

Introduction

The use of medicinal plants to restore health has evolved over time from the simplest forms of local treatment, probably used by cavemen, to the technologically sophisticated forms used by the pharmaceutical industry in modern times. In spite of the enormous differences between the two forms of use, there is a common fact between them: in both cases, man has perceived the existence of something that, when administered in the form of a complex mixture such as teas, bottles, tinctures, powders, etc., or as a pure substance in isolation and transformed into tablets, drops, ointments or capsules, has the property of causing beneficial reactions in the body, capable of resulting in the recovery of health and cooperating to improve the quality of life (LORENZI; MATOS, 2008).

The use of medicinal plants to prevent, treat and cure diseases is one of the oldest ways used by humanity (VEIGA JUNIOR et al., 2005). Historical records dating back to 4,000 years before Christ prove the use of plants in the search for a cure for illnesses by the Egyptian and Chinese peoples, among others. Indigenous peoples used them as a remedy for their ailments and as a poison in their wars and hunts (CARVALHO, 2004). Most of the time, the culture of medicinal plants is the only therapeutic tool of isolated populations and ethnic peoples (CALDERON et al., 2009).

Since 1977, the World Health Organization (WHO) has encouraged the study of plants traditionally known as medicinal, with the aim of scientifically evaluating the benefits of using herbal medicines and at the same time knowing the safety of their misuse (YUNES; CALIXTO, 2001).

Medicinal plant is "any plant that contains, in one or more organs, substances that can be used for therapeutic purposes or that are precursors of semi-synthetic drugs", according to the World Health Organization (WHO, 2002). Brazil has the greatest biodiversity in the world, representing approximately one third of the total. Most of it is found in the Amazon Rainforest and the Atlantic Rainforest (CARVALHO et al., 2007; MACEDO, 2009). According to the List of Species of the Flora of Brazil (2014), it has around 43,894 catalogued species.

The search for complementary therapeutic alternatives to traditional pharmacotherapy has recently become an important guideline in the context of public health. Currently, research into medicinal plants is the fundamental basis of the health policies implemented by the government as part of the National Policy for Integrative and Complementary Practices of the Unified Health System (BRASIL, 2006).

The correct use of a plant's active ingredients involves proper preparation, i.e. for each part to be used, active ingredient to be extracted or disease to be treated, there are appropriate forms of preparation and use. There are few adverse effects when using herbal medicines, as long as the correct dosage is used. Most of the known adverse effects are external to the method of preparation and are linked to various processing problems, such as incorrect identification of plants, lack of standardization, poor processing, contamination, adulteration and substitution of plants, incorrect preparation and/or dosage (ARNOUS; SANTOS; BEINNER, 2005).

Thanks to the activity of secondary metabolites in higher plants, which use them as a defense mechanism to combat predation by microorganisms, insects and herbivores, they produce substances capable of providing antibiotic actions (GONÇALVES *et al.*, 2005).

Phytotherapy is defined as therapy that uses medicinal plants in their various pharmaceutical preparations and does not use isolated active substances. This therapy is based on herbal medicines, which are medicines acquired solely from plant raw materials (NASCIMENTO et al., 2010). Brazilian public policies have been created for actions on medicinal plants and herbal medicines (BRASIL, 2010).

According to the National Health Surveillance Agency (ANVISA), antibiotics are drugs that have characteristic substances capable of banishing or hindering the proliferation of bacteria responsible for pathologies or harm to the body (BRASIL, 2010). From the 17th century onwards, antibacterial agents began to be used to treat infectious diseases, which must demonstrate selective toxicity, i.e. they must be effective in fighting the target bacteria (DAVIS, 1987; RANG; DALE; RITTER, 2001; BROOKS et al., 2008; SOFIATI, 2009).

Although there is a great diversity of antibiotics that act on various pathogenic microorganisms, studies are looking for the ideal antibiotic, i.e. the one with the broadest spectrum of action, the lowest cost, the least toxicity and the least evidence of bacterial resistance, given that there is already bacterial resistance to some antimicrobial products (PAZHANI, 2004). This desired antibacterial activity can be found in medicinal plant species. In a population with low access to medicines, such as Brazil's, adding scientific guarantees to this therapeutic practice brings many advantages (FURLAN, 1998).

There is a perceived interest in medicinal plants as a complementary therapy, which is why it is important to study new extracts of medicinal plants with antimicrobial activity, in order to obtain new products that show a greater spectrum of action, less toxicity, lower cost and, above all, that have a low level of bacterial resistance (NASCIMENTO et al., 2000; PAZHANI, 2004).

Medicinal plants and herbal medicine

Since the dawn of time, humans have looked to nature for ways to improve their living conditions. The use of medicinal plants has occurred, in addition to their food character, as a result of their medicinal properties, whether concrete or ritualistic (LORENZI; MATOS, 2008). Their use in the treatment of pathologies is related to folk medicine in various parts of the world (SARTORATTO et al., 2004).

Over time, human society has acquired a wealth of information about the environment in which it lives, making it possible to exchange information directly with the environment out of a need for survival. This collection includes knowledge about the plants with which individuals come into contact. In this way, the search for and use of plants with therapeutic potential is an activity that has been passed down from generation to generation, described with the aim of preserving this ancient tradition and reported in various herbal medicine treatises (CORREA JUNIOR, 1991).

The use of medicinal plants for phytotherapeutic purposes was based on empirical knowledge, obtained by observing domestic herbivorous animals or by the ingestion of this or that plant by the human organism itself (survival instinct) (CUNHA, 2003).

For a long time, the use of medicinal plants was the main therapeutic resource used to care for the health of people and their families; however, due to the advances that have taken place in the technical-scientific environment, especially in the field of health sciences, various ways of treating and curing illnesses have appeared. One of these is the use of industrialized medicines, which were gradually introduced into the daily lives of modern people through advertising campaigns promising to cure the most diverse diseases. Since then, the use of medicinal plants has gradually been replaced by allopathic medicines (ALVIN et al. 2004).

In this way, plants are used as the only therapeutic resource by a portion of Brazilians and more than two thirds of the world's population. The main factors influencing the continuation of this practice include the population's low standard of living and the high cost of medicines. In this way, plant users all over the world continue to consume phytotherapics, thus rendering valid some of the therapeutic information accumulated over centuries (NEWALL et al., 2002).

However, popular and even traditional use is insufficient to prove that medicinal plants are effective and safe medicines. In this respect, phytotherapics are compared to any other synthetic xenobiotic, and the indication or official authorization of their use as medicines must be based on experimental evidence that the benefits that may result outweigh the risks to which those who use them are exposed (BRASIL, 1995).

There has been a growth in phytotherapy in Brazil, making it an important economic sector. This growth is due to the great demand for crude extracts and essential oils as an alternative for curing infections, as synthetic products have become ineffective and expensive, as well as the search for products that are less aggressive to the body (LIMA et al., 2006; HARVEY, 2008).

In Brazil, the National Policy for Integrative and Complementary Practices (PNPIC) in the Unified Health System (SUS) was established by Ministry of Health Ordinance

No. 971 of May 3, 2006. The aim of this decree is to increase therapeutic options for SUS users, guaranteeing access to medicinal plants, herbal medicines and services related to herbal medicine, with safety, efficacy and quality, from the perspective of comprehensive health care. In addition, the National Medicinal Plants and Phytotherapics Program, established in 2007, aims to ensure that the Brazilian population can obtain the rational and safe use of medicinal plants and phytotherapics, favoring the sustainable use of biodiversity, the advancement of the production chain and the national industry. In order to achieve this programme, one of the proposals is to Promote and recognize popular and traditional practices for the use of medicinal plants, herbal medicines and home remedies (BADKE, 2011).

Plants are an important source of bioactive natural products, many of which form models for the synthesis of many drugs. These products represent an enormous diversity of structure and physico-chemical and biological properties (PINTO et al., 2002). Despite the large amount of research in this area, the data available shows that only 15 to 17% of plants have been investigated for their medicinal potential (SIMÔES et al., 2003).

A variety of new species are being discovered and added to the country's list of diversity . Tropical forests have an enormous biodiversity, contributing as they do to the production of medicines. Several plants from the Brazilian flora are being used as natural drugs to treat various diseases, such as leishmaniasis, malaria, fungal and bacterial infections, and there is a huge use of extracts for their biological activities *in vivo* and *in vitro* (REIS; MARIOT; STEENBOCK, 2004; GULLECE et al., 2006).

In view of the various conflicts and recognition of the medicinal power of plants by the scientific community, the potential of Brazilian flora for use in medicine has been highlighted by scientists interested in discovering new substances with therapeutic effects (MALAFAIA et al., 2006; BRASIL, 2009), prompting the Ministry of Health to publish a list of 71 medicinal plants, with the aim of directing studies so that these plants could be implemented as an alternative form of therapy for the population of the Unified Health System (ANVISA, 2011).

The Northeast region of Brazil has a great biodiversity in its ecosystem, with a predominance of Caatinga, a specific habitat for plants with medicinal and aromatic properties, considering it one of the richest sources of potentially pharmacological material in the world, as a result of the diversity of species and the knowledge derived from the traditional medicine of indigenous, black and European cultures (BRANDÂO et al., 2004).

Family Myrtaceae

The Myrtaceae family is found on all continents except Antarctica, but is more prevalent in the tropical and subtropical regions of the world (BARROSO 1991, MARCHIORI; SOBRAL, 1997).

In Brazil, the family is one of the most important in most vegetation formations (SOUZA; LORENZI, 2008), and is often the dominant family of shrubby tree species in the Dense Ombrophilous Forest (ASSIS 2004; MORI *et al.,* 1983). It is mainly distributed in the Atlantic Forest, where 636 species are found. It is considered one of the best represented families in Brazil (SOBRAL et al., 2009).

The family is divided into the subfamilies Leptospermoideae, where they are distributed in Oceania, and Myrtoideae, with a pantropical distribution and a center of irradiation in South America, and are mainly characterized by their dry and fleshy fruits, respectively (LEGRAND; KLEIN 1967; SOBRAL, 2003).

The majority of Brazilian myrtle species do not producc valuable wood, but are restricted to the supply of firewood and are used in small pieces or objects and other forms of local use (MARCHIORI; SOBRAL, 1997). On the other hand, there are numerous fruit species, some of which are widely exploited and commercialized (the guava tree, *Psidium guajava*, the jaboticaba tree, *Myrciaria cauliflor* O. Berg, and the pitangueira, *Eugenia uniflora*). These species represent a small fraction of the family's great economic potential, given that we have a large number of edible fruits produced by non-commercial species (LANDRUM; KAWASAKI, 1997).

According to Lorenzi and Souza (2001), for ornamental purposes, the "myrtle" species

(*Eugenia sprengelii* DC.) and the "erica" species (*Leptospermum scoparium* J.R.Forst; G.Forst.) stand out. As for the use of medicinal plants, *Eucalyptus globulus* L., used to treat flu, nasal congestion and sinusitis, and *Myrciaria dubia*, camu-camu, which has a high vitamin C content, according to ethnopharmacological information (LORENZI; MATOS, 2002). They may also be ecologically important, as their juicy, fleshy fruits are a source of food for wildlife. Most of the animals that feed on these fruits end up carrying and dispersing the seeds, favoring the survival and permanence of these species (PIZZO 2003; GRESSLER et al., 2006).

In the taxonomic system, Myrtaceae belongs to a group of plants whose division is Magnoliophyta, the class Magnoliopsida and the order Myrtales (ANGIOSPERM PHYLOGENY GROUP III, 2003). It is one of the most complex families from a taxonomic point of view, both in terms of the number of species and the scarcity of taxonomic studies (SOUSA; LORENZI, 2008).

Syzygium cumini

Syzygium cumini, a member of the Myrtaceae family, is popularly known as jambolao, black olive and purple plum. The plant was introduced to many tropical regions such as East and West India and Africa, as well as Brazil, and is found in several states in the southeast, northeast and north (DANADIO; SACRAMENTO, 1998). Its synonyms are *Myrtus cumini (*L.), *Eugenia jambolana* (Lam.), *Syzygium jambolanum* (Lam.), *Calyptranthes oneilli* Lundell and *Eugenia cumini*. The bark of this plant is used as a hypoglycemic medication to control diabetes. Phytochemical analysis of the fruit shows the presence of tannins, starch, gallic acid, phenolic resin and jambulol (LORENZI; MATOS, 2002).

The official or *pharmacopoeial* application of the fruit of *Syzygium cumini* is shown in *The British Herbal Pharmacopoeia*, 1983, in the form of a dry or hydroethanolic extract in a ratio of 1:1 in 25% alcohol. It is indicated for the treatment of diarrhea, and the other activities described by the same reference are: anti-inflammatory, astringent, carminative and hypoglycemic action. The recommended doses are 0.3 to 2.0 g of the dry fruit extract, by infusion or decoction, and 2 to 4 mL of the hydroethanolic extract in daily doses, three times a day (KETYLIN, et al., 2006).

According to Louguercio et al. (2005), the leaves of *Syzygium cumini* are rich in tannins and saponins. The bark, leaves and seeds are very astringent. The juice of the fruit is used as an astringent, diuretic, antidiabetic and stomachic. Boiling the bark is very effective as a mouthwash in the treatment of canker sores, stomatitis, sore throats and other oral diseases, leukorrhea, venereal ulcers, blood purification, stopping bleeding in the stool, dysentery, dyspepsia, asthma, bronchitis, burns, urinary retention and scaling of the scalp.

Tannins and other phenolic compounds are being recognized as having high antimicrobial potential. This has been proven by certifying the use of medicinal plants rich in polyphenols to obtain antimicrobial effects (KOLODZIEJ et al., 2003).

Eugenia uniflora

Eugenia uniflora, popularly known as pitanga, belongs to the Myrtaceae family and is native to Brazil, found from the Southern Plateau to the coastal areas of the Northeast to the South (LORENZI, 2008). As for the use of this plant in folk medicine, its efficacy and safety have yet to be proven, but it continues to be used as the basis of popular tradition, being considered an excitant, febrifuge, antirheumatic and antidysenteric (ALBUQUERQUE, 1989). Pharmacological tests on extracts of this plant have shown inhibitory activity of the enzyme xanthine oxidase due to the presence of flavonoids and antimicrobial activity against some pathogenic germs (LORENZI, 2008).

The use of *Eugenia uniflora* leaves, whose infusion has been used in folk medicine as an anti-hypertensive and anti-rheumatic agent, is well recognized in Brazil (LORENZI; MATOS, 2002).

Eugenia uniflora (pitangueira) is known as a semi-deciduous shrub or tree, 4 to 10 m high, with a narrow crown and a smooth, light brown trunk. The leaves are simple, chartreuse, 3 to 7 cm long, with a characteristic smell. The flowers are white, single or in clusters. The fruits are drupe-like, globose and grooved, shiny and reddish, yellowish or black in color, with fleshy, bittersweet pulp, containing 1 to 2 seeds (LORENZI;

MATOS, 2002).

Eugenia uniflora was used in empirical medicine by the Guarani Indians in the 15th century (ALONSO, 1998) and is used as an antihypertensive, diuretic (CONSOLINI; SARUBBIO, 2002), astringent (BANDONI et al., 1972), antipyretic and for the treatment of digestive diseases (ALICE et al., 1991). On the island of Madeira, the leaves of *Eugenia uniflora* are used to treat bronchitis, flu and intestinal problems and in Nigeria as a febrifuge (CONSOLINI; SARUBBIO, 2002).

According to Schapoval et al. (1994) the leaves of *Eugenia uniflora* are rich in essential oils containing citronellol, geraniol, cineol and sesquiterpenes which have been shown to have antimicrobial activity.

Syzigium malaccense

Syzygium malaccense also belongs to the Myrtaceae family and is popularly known as jambo-vermelho. In Brazil, it is found in the states of the North, Northeast and hot regions of the Southeast. It is a plant with edible fruits and its leaves are used in folk medicine as a diuretic and in the treatment of skin infections, gastrointestinal tract infections, respiratory infections and inflammations, as it inhibits the action of COX-1 during the biosynthesis of prostagladins (DUNSTAN et al., 1997; NOREEN et al., 1998).

Syzigium malaccense can reach heights of 12 to 15 m. It has a straight trunk and a dense, pyramidal crown with abundant branching that begins 1.5-2.0 m above the ground (CAVALCANTE, 1996). From August to February, the flowering season, it is covered in reddish flowers, which characterizes its ornamental appearance. The fruit begins to develop from this time until harvest time, which is from January to May. Propagation usually takes place by seed (DANADIO, 1998).

The fruit is usually eaten *fresh* or in the form of jams. According to Lorenzi et al. (2006), they contain vitamins A, B1 and B12, as well as calcium, iron and phosphorus. According to Donadio et al. (1998), the pulp makes up 84% of the fruit and has a Brix of 6.8% and an acidity of 0.4% at the end of ripening (AHMAD; ISMAIL, 2003).

According to Noreen et al. (1998), phytochemical studies on the leaves of *Syzygium malaccense* showed four flavonoids: catechin, mearnsitrin, myricitrin and quercitrin.

Locher and colleagues (1995) evaluated the antimicrobial and anti-inflammatory activity of some Hawaiian medicinal plant extracts. The authors found that the aqueous extract of the stem bark of *Syzygium malaccense* at a concentration of 125 µg/mL inhibited the growth of the Herpes Simplex-1 virus. They also studied the antibacterial action of the aqueous extract of *Syzygium malaccense* leaves by the method of diffusion in solid medium and observed an inhibition of the growth of strains *of Streptococcus pyogenes* and *Staphylococcus aureus* at a concentration of 100 µg/mL. Inhibition of the classical complement pathway and activation of the alternative complement fixation pathway were also observed with the acetonitrile extract obtained from the stem bark of *Syzygium malaccense.*

Psidium guajava

Psidium guajava, popularly known as the guava tree, grows in the wild as an evergreen shrub of the Myrtaceae family. It is a fruit tree from Central and South America and can be cultivated in all countries with a tropical climate (VENDRUSCOLO et al., 2005; TÔRRES et al., 2005). Its leaves have the following chemical composition: tannins (9-10%), essential oil (90.3%) rich in caryophyllene, nerolidiol, β-bisabolene, aromadendrene, *p-selinem, α-pinem* and 1,8- cineol; triterpenoids (oleanolic, ursolic, catecholic, guaiavolic, maslinic acid), β-sitosterol. The stem contains 12 to 30% tannins (ALONSO, 1998).

Psidium guajava, as well as being a popular fruit in tropical and subtropical countries, is widely used in folk medicine. Extracts and metabolites from the plant, especially those from the leaves and fruit, have useful pharmacological actions (GUTIÉRREZ; MITCHELL; SOLIS, 2008).

This species is known to be used mainly against colic and diarrhea, and also has a diuretic action. It also has antimicrobial and hypoglycemic activity (SANTOS et al.,1998Several authors have evaluated the activity of the leaf's aqueous extract against *Pseudomonas aeruginosa, Escherichia coli, Sarcina lutea, Serratia marcescens,*

Shigella flexneri, Staphylococcus albus and *Staphylococcus aureus* (CACERES et al., 1993).

Family Lamiaceae

The Lamiaceae family, formerly Labiatae, has approximately 240 genera and around 7,200 species, the vast majority of which are aromatic and have enormous chemical variety (BRAUCHLER; MEIMBERG; HEUBL, 2010; DO et al., 2014). It has the most widely used medicinal plants in the world and is mainly native to the Mediterranean area, although some originated in Australia, South-West Asia and South America (CUPPETT; HALL, 1998).

The species are herbaceous, shrubby or sub-shrubby and can be annual or perennial, with simple or compound crossed or opposite leaves, which can be sessile or petiolate. The flowers are bilabiate, of different sizes, alone or together in a dense terminal or axillary inflorescence and very aromatic (AGOSTINI et al., 2009). They grow in various habitats and altitudes, from the Arctic region to the Himalayas, Southeast Asia to Hawaii, Australia, Africa and from the North to the South of the New World. They are a typical Cerrado family, inhabiting the hottest regions of the world (ALMEIDA; ALBUQUERQUE, 2002).

It is very useful in traditional medicine, since several of its species are used in the therapy of tumors, swellings, dermatitis, skin irritations, colic, and used as antiparasitics, antimicrobials, among others (HARLEY; REYNOLDS, 1992). They are also used as aromatics in the cosmetics, food, hygiene and perfumery industries (AGOSTINI et al., 2009).

According to bibliographical research, the great use of this family in medicine is associated with terpenoids, especially those found in essential oils, which together with flavonoids are considered the most abundant metabolites found in the Lamiaceae family (AGOSTINI et al., 2009).

Ocimum gratissimum

The species *Ocimum gratissimum* Lineu (Figure 1), popularly known as alfavaca,

alfavaca cravo or alfavacào, is an odoriferous, straight sub-shrub from Asia and Africa and subspontaneous throughout Brazil (LORENZI; MATOS, 2002).

Figura 1 - Leaves and seeds of the *Ocimum gratissimum* species

Its leaves are used in folk medicine to make antiflu baths, especially for children, and to treat nervous conditions and paralysis. It is also used in teas as a carminative, sudorific and diuretic (MATOS, 2000).

According to studies, the essential oil in the leaves is about 3.60% eugenol, 77.3% eugenol, 12.1% 1.8-cineole, 2.3% b-caryophyllene and 2.1% (Z)-ociment (MORS; RINIZZI, 1966; SILVA, 1996). Research has shown that the highest eugenol content occurs between 12 noon and 5 p.m., while 1,8-cineole, an aromatic principle with an antiseptic and expectorant action, is highest at the end of the day and in the morning (SILVA, 1996; MATOS, 2000).

The essential oil found in the leaves of this plant has a bactericidal action and has been found to be bioactive against highly virulent organisms such as *Staphylococcus aureus*, *Bacillus spp*, *Pseudomonas aeruginosae*, *Klebisiella pneumoniae*, *Proteus mirabilis* and *Leishmania amazonensis* (UEDA-NAKAMURA et al., 2006; MATASYOH et al., 2007).

Plectranthus amboinicus

The species *Plectranthus amboinicus* (Figure 2) is a large, tall, very balsamic, semi-fleshy herb with deltoid-oval leaves, a toothed edge and prominent veins on the back

(ROBINEAU, 1995; MATOS, 2002). Popularly known as Malvariço, thick-leaf mint (ABDEL-MOGIB et al., 2002).

Figura 2 - Leaves of the species Plectranthus amboinicus

It comes from the island of Amboino in New Guinea and has been planted in tropical and subtropical countries, including Brazil, and is used in home gardens for medicinal purposes (MATOS, 2002).

According to ethnopharmacological knowledge, the leaves of this species are used to prepare homemade syrups to treat sore throats, coughs and bronchitis, and are also known as holy mallow lickers, as well as being used to treat wounds caused by cutaneous leishmaniasis. The juice of the leaves is used orally for ovarian and uterine problems, and even in cases of cervicitis (MATOS, 2002).

Phytochemical studies show the presence of essential oils abundant in thymol and sesquiterpenes, mucilage, quercetin, luteolin and other flavonoids and some triterpenic acids (BRIESKORN; RIEDEL, 1977; ROBINEAU, 1995; VASQUEZ et al., 1998). This species has been used as an anti-inflammatory therapeutic and also for coughs and bronchitis. However, considering the presence of thymol in its essential oil, the use of the leaves as a mouthwash and wound therapy has been proven, and it has antimicrobial activity (MATOS, 2002; MOREIRA et al., 2002).

Plectranthus barbatus

The species *Plectranthus barbatus* (Figure 3) comes from Africa and is an aromatic shrub with velvety, flexible leaves with toothed margins. It is popularly known as mallow and boldo (ALBUQUERQUE, 2000). It is one of the most mentioned plants in ethnobotanical studies of medicinal plants in Brazil (COSTA; NASCIMENTO, 2003).

Figura 3 - Leaves of the species *Plectranthus barbatus*

According to pharmacological analyses carried out at the Escola Paulista de Medicina, the aqueous extract has a predominantly gastric hyposecretory action, and is indicated for gastritis, heartburn and dyspepsia. In addition to this, various pharmacological actions have been tested and verified, including bronchodilator, cardiovascular and hypotensive actions, among others. Its pharmacological peculiarities and chemical components are not completely clear, but it is known that guaiene and fenchonsa belong to the essential oil of its leaves and are responsible for the essence (MATOS 2007).

Mentha arvensis

The *Mentha* genus *is* made up of 25 to 30 species and is distributed on almost all continents, standing out for its culinary and medicinal properties. They are recognized for the properties of their constituents and products that incorporate great financial relevance with the use of volatile and aromatic essential oils widely used by the food, cosmetic and pharmaceutical industries (BASER et al., 1999).

Mentha arvensis (Figure 4), known as Japanese mint, is a perennial plant with oval-shaped, shallowly toothed leaves. It originated in the Orient and was brought from Japan by immigrants (LORENZI, MATOS, 2002).

Figura 4 - Leaves of the *Mentha arvensis* species

It is an aromatic, stimulating and carminative plant. Its essential oil is rich in menthol, which is responsible for the smell of the plant and for its pharmacological actions, including nasal decongestant and tea for gastrointestinal problems. They are widely used to improve the smell of pharmaceutical preparations (LORENZI; MATOS, 2002).

Family Anacardiaceae

This botanical family includes trees, shrubs, lianas and rarely perennial herbs. The species are mainly distributed in the tropics. The main genera in this botanical family are *Anacardium* and *Mangifera* (Anacardiae), *Spondias, Lanneae* and *Tapirira* (Spondiadeae), *Semecarpus* (Semecarpeae), *Schinus, Rhus* and *Ozoroa* (Rhoeae) and *Dobinea* (Dobineae). Of these genera, *Pistacia lentiscus, Pistacia terebinthus* and *Rhus coriaria* stand out in temperate countries, while in Brazil the main genera are *Anacardium, Mangifera, Spondias* and *Schinus* (DI STASI; HIRUMA-LIMA, 2002).

Many species in this family produce fruits that are highly prized throughout the world, such as cashew, mango and seriguela, while others are important sources of wood. In Brazil, many species are spread throughout the territory, some with a wide occurrence in the Northeast Region, especially as a source of widely consumed and marketed fruit. Many of these species are used as medicinal plants in various regions of the country. In the Amazon region, the species *Anacardium occidentale* (cashew), *Anacardium giganteum* (moranha) and *Spondias purpurea* (seriguela) have been widely used (DI

STASI; HIRUMA-LIMA, 2002).

Anacardium occidentale

This species is widely known as Cajueiro, or simply Caju, in reference to the name of its fruit; however, various other names are used for the species, such as acajaiba, acajuiba, cajumanso, caju-manteiga, caju-da-praia, caju-de-casa, among others (DI STASI; HIRUMA-LIMA, 2002).

The juice of the fruit is used as a soft drink, and dried and roasted chestnuts are very popular all over the world. Economically, the nut oil, like the nut itself, has an important national and international market. The oil is used in the production of rubber, plastics and resins (DI STASI; HIRUMA-LIMA, 2002).

Anacardium occidentale (cashew) is a tree from the Anacardiaceae family. This botanical family contains species used by humans for food, timber and medicinal purposes, such as aroeira, caju, cajà, mango and seriguela. Low canopy, 5-10 m high. Simple leaves, 8-14 cm long. In home medicine, oral preparations are used, made with the shell, the gum and the cashew nut liquid (LCC), traditionally considered to be antidiabetic, astringent, antidiarrheal, purifying, tonic and antiasthmatic (LORENZI; MATOS, 2008).

Cashew buds are used against stomach aches and digestive problems and should be boiled with guava buds (DI STASI; HIRUMA-LIMA, 2002).

Spondias purpurea

The species is known in the Amazon region as umbu or seriguela. Numerous species of this genus are historically known as Cajazeiro and umbuzeiro, or cajà and umbu, referring only to the fruit of the species. Other common names are acajà, acaju, acaiou, cirouela, siriuela, or even caju (DI STASI; HIRUMA-LIMA, 2002).

This is a deciduous tree 3 to 6 meters high, with discreet flowers, pinnate leaves and drupe-like fruits that can be found in green, yellow, orange, red and violet. The fruit can reach 5.5 cm and weigh between 12-28 g and can taste bitter, sweet or sour. The adult plant rarely exceeds 7 m in height. It is difficult to propagate by seeds and its multiplication, due to anthropogenic action, is by cuttings. Fruiting begins in the third year after planting and can produce between 80 and 120 kilos a year (FREIRE, 2001).

Brito (2010) describes that the extract of the leaves and bark of the seriguela is used as an antipyretic and antidiarrheal, as well as in the treatment of gum infections, skin rashes and measles. The plant's essential oil contains terpenes that indicate a possible antimicrobial action. Research into the antimicrobial action of this plant is limited.

Mangifera indica

The mango tree *(Mangifera indica) is* a fruit-bearing species of the Anacardeaceae family that originated in India, where it spread to many regions with a tropical climate. In Brazil, the main production centers are located on the banks of the São Francisco River, although other states also grow the species. The main producers of the fruit in the country are the states of Sao Paulo and Minas Gerais, which together account for around 50% of the planted area and 25% of total production. This is followed by the northeastern states, which account for half of the country's production, especially Bahia, Pernambuco, Piaui and Cearà. The state of Bahia is the largest producer in the Northeast with 37.71% and a harvested area of 10,662 hectares (AGRIANUAL, 2002).

According to Panizza (1997), the most commonly used parts of this plant for therapeutic purposes are the leaves, fruit and bark of the branches, which are indicated for lung diseases, inflamed gums and decubitus ulcers.

Antibiotics

Antimicrobial or antibiotic substances are a special group of therapeutic agents, generally produced and obtained from living organisms. They are substances which, in small concentrations, should possess lethal or inhibitory activity against many microbial species, and, in addition to preventing the development of resistant microorganisms, should be free of undesirable effects on the host and chemically stable, among other characteristics (COWAN, 1999).

Antibiotics are the result of natural metabolites from fungi, actinobacteria and bacteria, capable of preventing growth or destroying microorganisms. Among the most common problems limiting the use of these antibiotics are the side effects produced in the body and the increase in microbial resistance (MIMS et al., 2005; BLACK, 2002).

The essence of antimicrobial chemotherapy is to kill or inhibit the microorganism without harming the host. Antimicrobial drugs can be classified as bacteriostatic or bactericidal, when they kill the microorganism or prevent its growth (TORTORA et al., 2012; ALTERTHUM, 2004).

Bacterial Diseases and Antibiotic Therapy

Bacteria are single-celled microorganisms with a simple structure and are present in a variety of places including the skin, mouth, respiratory tract, intestines and genital organs. They hardly cause infections unless the body's defenses are weakened (BURTON, 2012).

Most infections are caused by pathogenic bacteria, which invade the body and multiply, using toxins, enzymes and virulence factors to affect cells and cause disease. They can only be seen under a microscope and measure around 0.5 to 1.0 microns (1/1000 mm). In general, they are composed of a cell wall, cytoplasmic membrane, cytoplasm, flagella, fimbriae, spores, capsule and nucleus (BURTON, 2012).

The bacteria used in this research are *Streptococcus pyogenes, Enterecoccus faecalis, Staphylococus aureus* and *Pseudomonas aeruginosa.*

Streptococcus pyogenes or beta-hemolytic streptococcus belonging to Lancefield's group A is the most frequent bacterial species associated with the etiology of infections primary of the pharynx and tonsils (CUNNINGHAM, 2000). More serious pathologies are those caused by toxins, such as scarlet fever and toxic shock syndrome. Sequelae related to infections caused by *Streptococcus pyogenes* include rheumatic fever, reactive arthritis, glomerulonephritis and erythema nodosum (EFSTRATIOU, 2000; STEER *et al.*, 2007). These infections occasionally occur in any age group, but are common in children and adolescents aged between 5 and 15 (BISNO et al., 1997). The incubation period for streptococcal pharyngotonsillitis is usually 12 to 24 hours. The patient may have a high fever and pain on swallowing, with general malaise, anorexia and asthenia, symptoms which in children in particular may be followed by nausea, vomiting and abdominal pain (BALDY, 1997).

Even though they are often detected as human pathogens, some strains are only asymptomatic colonizers. Streptococci can be isolated from food, respiratory and genital tract microbiota and colonize the skin and mucous membranes (BALDY, 2007).

Enterococcus faecalis are gram-positive cocci, facultative anaerobes and can be found isolated, in pairs or in short chains. This microorganism is rarely found in primary infections, but it is the species that predominates in cases of endodontic retreatment, making up 38 to 70% of the microbiota in these cases, and is found in twenty out of thirty cases of persistent endodontic infections associated with teeth with filled canals. This high prevalence in complications following endodontic treatment is associated, in particular, with its ability to adhere to collagen, enhanced by the presence of human serum and, in the process of colonization, to invade the dentinal tubules, in addition to influencing the host's defenses and resisting the action of antimicrobial substances, making its colonization a source of constant reinfection of the filled canal (SÓ et al., 2011).

From a microbiological point of view, enterococci have low growth requirements, are able to grow at a temperature of 10 to 45 °C, pH 9.6 in 6.5% saline solution, and survive at 60 °C for 30 minutes. They are facultative, catalase-negative microorganisms, and a large proportion of the species in this genus hydrolyze esculin in the presence of bile (KONEMAN et al., 1997).

The bacterium called *Staphylococcus aureus* is a spherical, gram-positive coccus, often found on the skin and in the nasal passages of healthy individuals. However, it can cause diseases ranging from simple infections such as pimples, boils and cellulitis to serious infections (pneumonia, meningitis, endocarditis, toxic shock syndrome, septicemia and others) (SANTOS et al. 2007).

Nasal colonization by *Sthaphylococcus aureus* is asymptomatic, i.e. the individual does not develop an infection. This symptomless colonization is of great clinical importance since, with colonized nostrils, the individual contaminates their own hands and becomes a vehicle for transferring the bacteria in the mechanism of contact infections. Thus, especially in hospitals, the asymptomatic host can be a patient, a visitor or even a healthcare professional. According to some studies, nasal carriage also contributes to the transmission of the bacteria by airborne dissemination (VELÂZQUEZ-MEZA, 2005).

The distribution of *Sthaphylococcus aureus* is quite wide, as this bacterium is able to resist desiccation and cold, and can remain viable for long periods in dust particles. This bacterium can be found in the human circulatory environment, with humans being its main reservoir, as well as being present in various parts of the body, such as the nasal cavities, throat, intestines and skin. Of these anatomical sites, the nostrils have the highest colonization rate, the prevalence of which is around 40% in the adult population, and may be even higher in hospital environments (HANNERMAN, 2003; CARVALHO et al., 2005; CAVALCANTI, 2005 *apud* SANTOS et al., 2007).

Pseudomonas aeruginosa is a pertercent member of the Pseudomonadaceae family. It is characterized as a straight or slightly curved Gram-negative rod, strictly aerobic, and can be found as single cells, in pairs, or in short chains, showing mobility through a monotrique polar flagellum (PALLERONI et al., 1998). Pyocyanin and pyoverdin are diffusible fluorescent pigments produced by this bacterium in the culture medium. *Pseudomonas aeruginosa* is a non-carbohydrate fermenter, produces 3 cytochrome oxidase, uses nitrate instead of oxygen as the final electron acceptor, and also produces Arginine hydrolase and Ornithine decarboxylase (JIMENEZ-ESCRIG, 2001).

The diseases caused by this agent range from superficial skin infections to fulminant sepsis (MURRAY, 2006). *Pseudomonas aeruginosa* can cause acute infection through the production of toxins and chronic infection due to the action of the thick layer that consists of its biofilm, and can also result in the sum of the types of infection through the concomitant action of these components (PALLERONI, 1998).

A striking characteristic of *Pseudomonas aeruginosa* infections caused in ICUs is multidrug resistance. The percentages of resistance are much higher in the samples isolated in these treatment units, which reflects the greater intensity of antimicrobial use in this environment, and probably the transmission of multidrug-resistant strains between patients (SADER et al., 1994; GOLDMANN; HUSKINS, 1997).

Antimicrobials are compounds that have the ability to inhibit the growth and/or kill microorganisms, they can be generated by bacteria or fungi or they can be totally or partially synthetic. The main purpose of using an antimicrobial is to prevent or treat an infection by mitigating or destroying pathogenic microorganisms and, if possible, preserving the microorganisms of the normal microbiota (MELO et al., 2012). They can be divided into bactericidal, when they cause the death of the bacteria, or bacteriostatic, when they cause the inhibition of microbial development (WALSH, 2003).

Antimicrobial or antibiotic substances form a special group of therapeutic agents, probably produced and obtained from living organisms. They are substances which, in small concentrations, generally have lethal or inhibitory activity against many microbial species and, in addition to preventing the development of resistant micro-organisms, must be free of adverse effects on the host and chemically stable, among other characteristics (COWAN, 1999).

Antimicrobials come from the natural metabolites of fungi, actinobacteria and bacteria and are capable of inhibiting growth or destroying microorganisms. The most common problems affecting the use of these antibiotics are the adverse effects produced in the body and the growing increase in microbial resistance (MIMS et al., 2005; BLACK, 2002).

Antibiotics are among the most significant health discoveries of the 20th century, changing the clinical outcome of various diseases and thus minimizing mortality. However, it is one of the classes of drugs most often prescribed and used indiscriminately, and as a result there is an increase in bacterial resistance (DURANTE- MANGONI et al., 2009).

A lot of research and studies have been carried out in search of antimicrobial properties and thus obtaining positive results. This research is carried out through *in vitro* biological assays, where these are simulated, analyzed and verified. These studies can be carried out with the aim of finding a new antimicrobial or defining the spectrum of action of a new drug, as well as investigating the resistance or sensitivity of bacteria and fungi (VOLPATO, 2005).

Among the widely used antibiotics is Cefuroxime, a second-generation cephalosporin synthesized from cephalosporin C. The minimum inhibitory concentration (MIC) of this antibiotic is around 4 µg/mL for most sensitive bacteria. However, depending on the bacterial strain and species, higher concentrations may be necessary to achieve clinical efficacy (PERRY; BROGDEN, 1996).

Among the various classes of existing antimicrobials, one represents a huge number of innovations in antibiotic therapy: cephalosporins (PICHICHERO; CASEY, 2007; JONES et al. 2007). Cephalosporins are considered to be of great importance, as they have been shown to be effective in the treatment of numerous infections, such as pharyngitis, pneumonia, otitis, urinary infections, skin infections, among others, and are also widely used in prophylaxis and post-surgery (PONG; BRADLEY, 2005; MAZZEI et al. 2006). They are classified by generation, each consisting of general characteristics of antimicrobial activity (HARDMAM; LIMBIRD, 2006). Of the second generation cephalosporins, cefuroxime has a broad spectrum of action, well-established pharmacokinetics and is designed for infections by gram-positive and gram-negative bacteria. It is resistant to β-lactamase and inhibits the synthesis of the bacterial wall and excites autolytic enzymes (MCEVOY, 2003).

The mechanisms of action can be divided into four categories: inhibition of cell wall synthesis; inhibition of cell membrane function; inhibition of protein synthesis; inhibition of nucleic acid synthesis, although the mechanism of action of most antimicrobials is not completely elucidated (TORTORA; FUNKE; CASE, 2003; BLACK, 2002; TRABULSI; ALTERTHUM, 2008; MADIGAN, et al., 2010).

Naturally occurring antibiotics and their semi-synthetic derivatives cover the majority of antibiotics in clinical use and can be classified as β-lactams (penicillins, cephalosporins, carbapenins, oxapenins and monobactams), tetracyclines, aminoglycosides, macrolides, cyclic peptides (glycopeptides, Iipodepsipeptides), streptogramins, among others (lincosamides, chloramphenicol, rifamycins etc). Antibiotics of synthetic origin are classified as sulfonamides, fluoroquinolones and oxazolidinones (ABRAHAM, 2003; PATRICK, 2005; PUPO et al., 2006).

Bacterial Resistance

Worldwide, there are more and more resistant bacteria and an increasing number of multidrug-resistant strains (particularly in hospital settings), increasing morbidity, essential healthcare costs and mortality rates from infections.

Since ancient times it has been known that man and microbes share a common life that is lost in the shadows of time; and certainly since prehistoric times they have been causing illness in man (OLIVEIRA; LIMA; MOURA, 2011).

In relation to bacteria, Rodrigues (1997) and Moura (2004) argue that, with the event of bacterial resistance to antibiotics, multidrug-resistant strains began to emerge, which are difficult to treat, requiring health professionals to fully adhere to standard precautions and isolation measures when caring for patients, avoiding their dispersal.

The incorrect use of antimicrobials is primarily responsible for the phenomenon of bacterial resistance, which is characterized by the ability of microorganisms, especially bacteria, to resist the lethal action of one or more classes of antibiotics (OLIVEIRA; PAULA, 2012). In addition, *there is* a growing use of new generations of antibiotics and an abandonment of older antibiotics (DIAS; MONTEIRO; MENEZES, 2010).

Bacteria are divided into resistant or sensitive according to Minimum Inhibitory Concentration (MIC) and Minimum Bactericidal Concentration (MBC) data. They are said to be resistant when they are inhibited *in vitro* only at concentrations higher than those reached *in vivo*. The resistance of a microorganism to a certain drug can be distributed primarily as intrinsic or acquired. Intrinsic resistance is that which participates in the natural, phenotypic properties of the microorganism, propagated only vertically to offspring (LIVERMORE, 1994).

The growing clinical relevance of hospital and community bacterial infections caused by multi-resistant strains is alarming, as is the ineffectiveness of the most commonly used antibiotics, prompting studies into the bioactivity of natural products, with an emphasis on essential oils and phytoconstituents isolated from medicinal plants (BENKEBLIA, 2004).

Before the 21st century, bacterial resistance occurred mainly in hospitals. Nowadays, bacterial resistance is associated with various locations and can reach healthy people. The study of the biochemical and genetic mechanisms involved in bacterial resistance is of great importance in understanding how bacteria can create resistance. Although these mechanisms vary from pathogen to pathogen (DEMAIN, 2002; LEVY et al., 2001; WOODFORD, 2005).

Resistance to antibiotics is a genetic event, associated with the existence of genes present in the microorganism that encode different biochemical mechanisms that prevent the action of the drugs. The indiscriminate use of antimicrobials is a decisive element in the emergence of multidrug-resistant microorganisms and, consequently, in the rise in hospitalization costs (WRIGHT, 2005).

While microbial resistance has been growing significantly, the amount of research into new antibiotics has fallen dramatically in recent years (COUTINHO, 2009).

The essential conditions that have led to the inappropriate use of antimicrobials are: ignorance of infectious diseases, doubtful diagnosis and a lack of awareness among professionals about the severity of the complication that is bacterial resistance (BRASIL, 2004; ANDRADE; LEOPOLDO; HAAS, 2006).

Methods can be applied to prevent the spread of antimicrobial resistance, including: the prevention of bacterial infections through the use of vaccines, the rational use of antibiotics, the management and prevention of the spread of resistant microorganisms, and the development and creation of new antibiotics (GUIMARÂES; MOMESSO; PUPO, 2010).

On the economic side, they are also worrying. Pharmaceutical companies aim to acquire a financial return over the course of development once they have an antibiotic on the market. However, the increased use of antimicrobials also forms bacterial resistance more quickly, which contributes to a reduction in patent time (FERNANDES, 2006).

A major concern about antimicrobial resistance is the increase and the future use of antimicrobials is undetermined. Measures should be taken to try to reduce this, such as: developing more research to try to better understand the mechanisms of bacterial resistance and rationalizing the use of antimicrobials (NASCIMENTO et al., 2000; HAIDA et al., 2007).

In order to control the use of antimicrobials, the Health Surveillance Agency (ANVISA), through Collegiate Board Resolution (RDC) No. 44 of October 26, 2010, imposes stricter rules on the packaging, labeling, release and control of medicines based on substances classified as antimicrobials (list of antimicrobials registered with ANVISA), for use on prescription, alone or in combination. Under this RDC, antimicrobials can only be dispensed in the presence of a doctor's prescription, in two copies, with the second copy being retained at the pharmacy (BRASIL, 2010).

It has become essential to look for new antimicrobial classes, to try to find new elements with antimicrobial properties to be studied in an attempt to combat resistant microorganisms, which is a major challenge. As a result, plants are sources of active substances with diverse functions, helping to advance the discovery of new drugs (SANDES; DIBLASI, 2000; CATÂO et al., 2005; HAIDA et al., 2007).

Brazil, due to its large and rich plant biodiversity, has great potential for discovering and developing new natural molecules with antimicrobial activity. Many plants are used as medicinals in therapies against various pathologies, including bacterial infections (ALVES et al., 2000).

The study of medicinal plants is an alternative in the search for new therapeutic agents (SANTOS, 2011).

According to Duarte (2006), the antibacterial properties that plants possess are the effects of the products of their secondary metabolism, substances such as 1,8-cineol, limonene and linalool, geranial, germacrene-D, and menthol have been recognized and confirmed scientifically. In this way, plant extracts and their phytochemical substances, with known antimicrobial properties, can be of great value in therapeutic treatments (NASCIMENTO et. al., 2000).

References

ABDEL-MOGIB, M. et al. Chemistry of the genus Plectranthus. **Molecules**, v. 7, n. 2, p. 271-301, 2002.

ABRAHAM, D. J. **Burger's medicinal chemistry and drug discovery**. WileyInterscience, 2003.

AGOSTINI, F. et al. Essential oil yield and composition of Lamiaceae species growing in southern Brazil. **Brazilian Archives of Biology and Technology**, v. 52, n. 2, p. 473-478, 2009.

ALBUQUERQUE, J.M. **Plantas Medicinais de Uso Popular**. Brasilia: ABEAS/MEC, 1989.

ALBUQUERQUE, R.L. Contribuiçao ao estudo quimico de plantas medicinais no Brasil: *Plectranthus barbatus* Andr *Plectranthus amboinicus* (Lour)Spreng.Fortaleza, 2000 166 p. (Dissertaçao de Mestrado) Universidade Federal do Cearà, Fortaleza, 2000.

ALICE, C. B. et al. Screening of plants used in south Brazilian folk medicine**. Journal of Ethnopharmacology**. v.35, p.165-171, 1991.

ALMEIDA, C.F.; ALBUQUERQUE, U.P. Check-list of the family Lamiaceae in Pernambuco, Brazil. **Brazilian Archives of Biology and Technology**, v. 45, n. 3, p. 343-353, 2002.

ALONSO T. **Tratado de Fitomedicina: bases clinicas y farmacológicas.** Buenos Aires: ISIS Ediciones SRL, 1998.

ALTERTHUM, F. **Origin and chemical nature of the main antibacterial agents**. Sao Paulo: Atheneu Publishing House, 2004.

ALVES, T. et al. Biological screening of Brazilian medicinal plants. **Memorial Instituto Oswaldo Cruz**, v. 96, n. 3, p. 367-373, 2000.

ALVIM, N. A. T. et al. **Technologies and techniques in health: how and why to use them in nursing care. Sao Paulo: Difusao Editora**, 2004.

ANDRADE, D.; LEOPOLDO, V.C.; HAAS, V. J. Occurrence of multiresistant bacteria in an intensive care unit of a Brazilian emergency hospital. **Revista Brasileira de Terapia Intensiva**, v. 18, n. 1, p. 27-33, 2006.

ARNOUS, A.H.; SANTOS A.S.; BEINNER, R.P.C. Medicinal plants for home use, popular knowledge and interest in community cultivation. **Revista Espaço para a Saùde**, v. 6, n. 2, p. 1-6, 2005.

BRIESKORN, C. H.; RIEDEL, W. Flavonoide aus Coleus amboinicus. **Planta Medica**, 1977.

BADKE, M.R. et al. Popular knowledge and practices of health care with the use of medicinal plants. **Texto & Contexto Enfermagem,** v. 21, n. 2, p. 363-370, 2012.

BARROSO, G.M. **Sistemàtica de Angiospermas do Brasil**.Viçosa: Universidade Federal de Viçosa, v. 2. 1991.

BASER, K. H. C. et al. Essential oils of Mentha species from Northern Turkey. **Journal of Essential Oil Research**, v. 11, n. 5, p. 579-588, 1999.

BENKEBLIA, N. Antimicrobial activity of essential oil extracts of various onions (*Allium cepa*) and garlic (*Allium sativum*). **Lebensmittel-Wissenchaft und-Technologie**, v. 37, n. 2, p. 263-268, 2004.

BISNO, A.L. et al. Diagnosis and management of group A streptococcal pharyngitis: a practice guideline. **Clinical Infectious Diseases** v. 25, p.574-83, 1997.

BLACK, J. G. **Microbiology: fundamentals and perspectives**. 4. ed. Rio de Janeiro: Guanabara Koogan, 2002.

BRANDÂO, M. G. L.; DINIZ, B. C.; MONTE_MÓR, R. L. M. Use of native species in popular medicine in areas of the Amazon. Medicinal plants: threatened knowledge. **Revista Ciência Hoje**, v. 35, n. 206, p. 64-66, 2004.

BRAUCHLER, C.; MEIMBERG, H.; HEUBL, G. Molecular phylogeny of Menthinae (Lamiaceae, Nepetoideae, Mentheae) - Taxonomy, biogeography and conflicts. **Molecular Phylogenetics and Evolution**, v. 55, n. 2, p. 501-523, 2010.

BRAZIL. National Health Surveillance Agency (ANVISA)**. Manual of Clinical Microbiology for Infection Control in Health Services.** Brasilia: Ministry of Health. 2004.

Available at: <www.anvisa.gov.br/servicosaude/microbiologia/introducao.pdf> Accessed on: 8 Sept. 2014.

BRAZIL. Ministry of Health. **National Program for Medicinal Plants and Herbal Medicines**. Brasilia, 2009.

BRAZIL. National Health Surveillance Agency (ANVISA)**. Antimicrobials - Theoretical Basis and Clinical Uses.** Brasilia: Ministry of Health. 2010. Available at:

<http://www.anvisa.gov.br/servicosaude/controle/rede_rm/cursos/rm_controle/opas_web/modulo1/conceitos.htm. Accessed on 22 Aug 2014.

BRAZIL, Ministry of Health, Health Surveillance Secretariat Ordinance No. 6/95 of 31.01.95. **Official Journal of the Union**, v. 200, section I, p. 1523, 6.2, 1995.

BRAZIL. Decree No. 5.813 of June 22, 2006. Approves the National Policy for Medicinal Plants and Herbal Medicines and makes other provisions. **Official Gazette of the Union**, Brasilia, DF, June 23, 2006.

BRITO, H. R. **Chemical characterization of essential oils from Spondias mombin, Spondias purpurea and Spondias sp (cajarana do sertao).** (Master's dissertation in Forestry Sciences). Federal University of Campina Grande, 2010.

BROOKS, G. F. et al. **Medical Microbiology**, 24 ed. Sao Paulo: Mcgraw Hill Interamericana do Brasil, 653p. 2008.

CALDERON, L. A. et al. Amazonian biodiversity: a view of drug development for leishmaniasis and malaria. **Journal of the Brazilian Chemical Society**, v. 20, n. 6, p. 1011-1023, 2009.

CACERES, Armando et al. Plants used in Guatemala for the treatment of gastrointestinal disorders. 3. Confirmation of activity against enterobacteria of 16 plants. **Journal of Ethnopharmacology**, v. 38, n. 1, p. 31-38, 1993.

CARVALHO, J. C. T. **Fitoteràpicos: anti-inflammatories: chemical and pharmacological aspects and therapeutic applications**. Tecmedd, 2004.

CARVALHO, A. C. B. et al. Aspects of legislation in the control of herbal medicines. **T&C Amazônia**, v. 5, n. 11, p. 26-32, 2007.

CARVALHO, C.E. et al. Sequential microbiological monitoring of tracheal secretion in intubated patients admitted to a pediatric intensive care unit. **Jornal de Pediatria**, v. 81, n. 1, p. 29-33, 2005.

CATÂO, R.M.R. et al. Evaluation of the antimicrobial activity of riparins on multidrug-resistant strains *of Staphylococcus aureus* and *Escherichia coli*. **RBAC**, v. 37, n. 4, p. 247-249, 2005.

CAVALCANTE, P.B. **Frutas comestiveis da Amazônia**. 6.ed. Belém: CNPq/ Museu Paraense Emilio Goeldi, 1996.

CONSOLINI, A. E.; SARUBBIO, M. G. Pharmacological effects of *Eugenia uniflora* (Myrtaceae) aqueous crude extract on rat' heart. **Journal of Ethnopharmacology**. v. 81, p. 57-63, 2002.

COSTA, M. C. C. D.; NASCIMENTO, S. C. Cytotoxic activity of *Plectranthus Barbatus* Andr. (Lamiaceae). **Acta Farmacêutica Bonaerense**, v. 22, n. 2, p. 155-158, 2003.

COUTINHO, H. D. M. **Avaliação da Atividade Antibacteriana e Fotossensibilizante de Produtos Naturais da Regiao do Cariri Cearense**. (Doctoral Thesis in Natural and Synthetic Bioactive Products). Federal University of Paraiba, 2009.

COWAN, M. M. Plant products as antimicrobial agents. **Clinical Microbiology Reviews**, v. 12, n.4, p. 564-582, 1999.

CUNHA, A. P. **Plantas e produtos vegetais em fitoterapia.** Lisbon: Caloste Gulbenkiam Foundation, 2003.

CUNNINGHAM, M.W. Pathogenesis of group A streptococcal infections. **Clinical Microbiology Reviews.** v. 2, n. 13, p. 470-511, 2000.

CUPPETT, S. L.; HALL, C. A. Antioxidant activity of the Labiatae. **Advances in food and nutrition research**, v. 42, p. 245-272, 1998.

DANADIO, L.C.; SACRAMENTO, V.K. **Exotic fruits**. Funep, Jaboticabal, 1998.

KETYLIN, F. MIGLIATO et al. Pharmacological action of *Syzygium cumini* (L.) Skeels. **Acta Farmacêutica Bonaerense**, v. 25, n. 2, p. 310-4, 2006.

DAVIS, B. D. Mechanism of bactericidal action of aminoglycosides. **Microbiological reviews**, v. 51, n. 3, p. 341, 1987.

DIAS, M.; MONTEIRO, M. S.; MENEZES, M. F. Antibiotics and bacterial resistance, old issues, new challenges, clinica farmacològica**, Clinica, investigaçao e inovaçao**,v. 3, n. 12, p. 3-11, 2010.

DEMAIN, A. L. Prescription for an ailing pharmaceutical industry. **Nature biotechnology**, v. 20, n. 4, p. 331-331, 2002.

DI STASI, L. C.; HIRUMA-LIMA, C. A. **Plantas medicinais na Amazônia e na Mata Atlàntica**. 2. ed. Sao Paulo: Editora Unesp, 2002.

DUARTE, M. C. Antimicrobial Activity of Medicinal and Aromatic Plants Used in the Brazil. **MultiCiência**, 2006. Available at: <http://www.multiciencia.unicamp.br/artigos_07/a_05_7.pdf>. Accessed on: November 25, 2015.

DUNSTAN, C.A *et al.* Evaluation of some Samoan and Peruvian medicinal plants by prostaglandin biosynthesis and rat ear oedema assays. **Journal of ethnopharmacology**, v 57, n.1, p. 35-56, 1997.

DURANTE-MANGONI, E. et al. Do we still need the aminoglycosides? **International journal of antimicrobial agents**, v. 33, n. 3, p. 201-205, 2009.

EFSTRATIOU, Androulla. Group A streptococci in the 1990s. **Journal of Antimicrobial Chemotherapy**, v. 45, n. suppl 1, p. 3-12, 2000.

FERNANDES, P. Antibacterial discovery and development-the failure of success? **Nature biotechnology**, v. 24, n. 12, p. 1497-1503, 2006.

FREIRE, F.C. O. Use of manipueira to control cerigueleira oidium: preliminary results. Technical communication**. EMBRAPA**, 2001.

FURLAN, M. R. **Cultivo de Plantas Medicinais.** Coleçao Agroindùstria, Cuiabà - Mato Grosso: 13.

ed. SEBRAE, 1998.

GOLDMANN, D. A.; HUSKINS, W. C. Control of nosocomial antimicrobial-resistant bacteria: a strategic priority for hospitals worldwide. **Clinical Infectious Diseases**, v. 24, p. S139-S145, 1997.

GONÇALVES, A. L.; ALVES FILHO, A.; MENEZES, H. Comparative study of the antimicrobial activity of extracts from some native trees, **Arquivos do Instituto Biològico**, v.72, n.3, p.353-358, 2005.

GUIMARÂES, D. O.; MOMESSO, L. D. S. E.; PUPO M. T. Antibiotics: Therapeutic Importance and Perspectives for the Discovery and Development of New Agents. **Quimica Nova**, v.33, n.3, 667-679, 2010.

GRESSLER, E. Pollination and seed dispersal in Myrtaceae from Brazil. **Revista Brasileira de Botànica**, v.29, n.4, p.509-530, 2006.

GULLUCE, M. et al. Screening the antioxidant and antimicrobial properties of the lichens Parmelia saxatilis, Platismatia glauca, Ramalina pollinaria, Ramalina polymorpha and Umbilicaria nylanderiana. **Phytomedicine**, v. 13, n. 7, p. 515-521, 2006.

GUTIÉRREZ, R.M.; MITCHELL, S.; SOLIS, R.V.J. *Psidium guajava*: a review of its traditional uses, phytochemistry and pharmacology. **Journal of Ethnopharmacol.,** Limerick, v. 17, no. 1, p. 1-27, Apr. 2008.

HAIDA, K. S. et al. In vitro evaluation of the antimicrobial activity of eight medicinal plant species. **Arquivos de Ciências da Saùde da Unipar, Umuarama**, v. 11, n. 3, p. 185-192, 2007.

HARDMAM, J.G; LIMBIRD, L.E. (Ed.) Goodmam & Gilmam: the pharmacological basis of therapeutics. 11th ed. New York: Mcgraw Hill, p. 1206-1213, 2006.

HARLEY, R.M.; REYNOLDS, T. Advances in Labiate science. **The Royal Botanic Gardens**, KEW, RICHMOND, UK. p.7-8, 455-468 1992.

HARVEY, Alan L. Natural products in drug discovery. **Drug discovery today**, v. 13, n. 19, p. 894901, 2008.

JIMÉNEZ-ESCRIG, A. et al. Guava fruit (Psidium guajava L.) as a new source of antioxidant dietary fiber. **Journal of Agricultural and Food Chemistry**, v. 49, n. 11, p. 5489-5493, 2001.

JONES, R. et al. Comparisons of parenteral broad-spectrum cephalosporins tested against bacterial isolates from pediatric patients: report from the SENTRY Antimicrobial Surveillance Program (1998-2004). **Diagnostic Microbiology and Infectious Disease**, v. 57, n.1, p. 109-106, 2007.

KONEMAN, E.W. et al. Color **atlas and textbook of diagnostic microbiology**. 5th ed. Chicago: Lippincott Williams & Wilkins; 1997.

LEVY, C. S. et al. Endocarditis caused by penicillin-resistant viridans streptococci: 2 cases and controversies in therapy. **Clinical Infectious Diseases**, v.33, n.4, p.577-579, 2001.

LIMA, M.R.F. et al. Anti-bacterial activity of some Brazilian medicinal plants. **Journal of Ethnopharmacology**, v.105, n. 1, p. 137-147, 2006.

LIVERMORE, D.M. Evolution of b-lactamase inhibitors. **Intesive Care Medicine**, v. 20, p. 10-13, 1994.

LORENZI, H.; MATOS, F.J.A. **Plantas medicinais no Brasil: nativas e exóticas cultivadas**, Sao Paulo: Instituto Plantarum, 2 ed. 2008.

LORENZI, H.; MATOS, F. J.A. **Plantas medicinais no Brasil: nativas e exóticas**. Instituto Plantarum de Estudos da Flora, 2002.

LANDRUM, L. R.; KAWASAKI, M. L. The genera of Myrtaceae in Brazil: an illustrated synoptic treatment and identification keys. **Brittonia**, v. 49, n. 4, p. 508- 536, 1997.

LOUGUERCIO, A. P et al. Antibacterial activity of hydro-alcoholic extract of jambolao leaves (*Syzygium cumini* (L.) Skells). **Ciencia Rural**, v. 35, n.2, p.373, 2005.

LOCHER, C.P. *et al.* Anti-microbial activity and anti-complement activity of extracts obtained from selected Hawaiian medicinal plants. **Journal of ethnopharmacology**, v. 49, n. 1, p. 23-32, 1995.

LORENZI, H. Arvores Brasileiras. **Manual of Identification and Cultivation of Native Brazilian Tree Plants**. 5. ed. Nova Odessa - SP: Instituto Plantarum, 2008.

LORENZI, H.; SOUZA, H. M. **Plantas Ornamentais no Brasil**. 3rd ed. Instituto Plantarum: Nova Odessa, 2001.

MACEDO, E.V. A produção de Fitomedicamentos e a Politica Nacional de Plantas Medicinais e Fitoteràpicos. **Revista Brasileira de Farmàcia**, v.90, n.4, p.290-97, 2009.

MALAFAIA, O. et al. Herbal medicines and their potential in surgical healing. **Acta Cirùrgica Brasileira**, v. 21, n. 3, 2006.

MARCHIORI, J.N.C.; SOBRAL, M. 1. **Dendrology of angiosperms - Myrtales**. Editora da UFSM, Santa Maria. 1997.

MATASYOH, L.G. et al. Chemical composition and antimicrobial activity of the essential oil of *Ocimum gratissimum* L. growing in Eastern Kenya. **African Journal of Biotechnology**, v.6, n.6, p.760-765, 2007.

MATOS, F.J.A. **Farmàcias Vivas - a system for using medicinal plants designed for small communities**. 4 ed. Fortaleza: Ediçoes UFC, 2002.

MCEVOY, G.K.Cephalosporins: cefuroxime sodium. Cefuroxime axetil. In: **AHFS Drug Information**. Bethesda: ASHP, p. 223- 231, 2003.

MIMS, C. et al. **Medical Microbiology**. 3. ed. Rio de Janeiro: Editora Elsevier, 2005.

MOREIRA, R.C.T. et al. Ethnobotanical approach to the use of medicinal plants in Vila Cachoeira, Ilhéus, Bahia, Brazil. **Acta Farmacêutica Bonaerense**, v. 21, n.3, p.1-7, 2002.

MORS, W.B.; RIZZINI, C.T. **Useful plants of Brazil.** San Francisco, HondenDay, 1966.

MOURA, J. P. **A adesâo dos profissionais de enfermagem às precauções de isolamento na assistência aos portadores de microrganismos multirresistentes**. School of Nursing, Ribeirao Preto. University of Sao Paulo, 2004. Available at:

<http://www.teses.usp.br/teses/disponiveis/22/22132/tde-12082004-125447/ptbr. .php> Accessed on: Nov. 24, 2015

MURRAY, P. R.; ROSENTHAL, K. S.; PFAÜER, M. A.. **Medical Microbiology**. 5. ed. Madrid: Elsiver Espana, 2006.

NASCIMENTO, J.J.M. et al. National Policy on Herbal Medicines. **Journal of**

Hospital administration and health innovation. v. 5, n. 4, p. 42-49, 2010.

NASCIMENTO, G.G.F. et al. Antibacterial activity of plant extracts and phytochemicals on antibiotic-resistant bacteria. **Brazilian Journal of Microbiology**, v.31, n.4, p.247-56, 2000.

NEWALL C. A.; ANDERSON, L. A.; PHILLIPSON, J. D. **Plantas Medicinas: Guia para profissional de saù**. Sao Paulo: Premier, 2002.

NOREEN, Y.; et al. Flavan-3-ols isolated from some medicinal plants inhibiting COX-1 and COX-2 catalyzed prostaglandin biosynthesis. **Planta Medica**, v. 6, n. 64, p. 520-524, 1998.

OLIVEIRA, F.B.M.; LIMA, L.M.; MOURA, E.B. **Indiscriminate use of antibiotics and microbial resistance:** a reflection on the treatment of hospital infections. 2011.

Available at :< uninovafapi.edu.br/sistemas/revistainterdisciplinar/v4n4/.../rev4_v4n4.pd...>.

Accessed on: 10 Aug 2015.

OLIVEIRA, A.C.; PAULA, A.O. Decolonization of Staphylococcus aureus carriers: indications, advantages and limitations. **Texto Contexto Enferm, Florianópolis**, v. 21, n. 2, p. 448-57, 2012.

WORLD HEALTH ORGANIZATION (WHS). **WHO strategy on traditional medicine** 2002-2005. Geneva 2002. 67p. <Available at: htpp://www.opas.org.br/ medicamentos/ site/ UploadArq/ trm_ strat_ spam.pdf.>. Accessed on: Aug 20, 2014.

PALLERONI, N. J. Introduction to the aerobic pseudomonads. In: Collier, L; Balows, A.; Sussman, M. (eds). Topley & Wilson's microbiology and microbial infections - **Systematic bacteriology.** v. 2, p. 1091-1108, 1998.

PANIZZA, S. **Plantas que curam**. 24. ed. Sao Paulo: Editora Ibrasa, 1997.

PATRICK, G. L.; **An Introduction to Medicinal Chemistry**, New York: Oxford University Press, ch.16, 2005.

PAZHANI, G.P. et al. Clonal multidrug-resistant Shigella dysenteriae Type 1 strains associated with epidemic and sporadic dysenteries in Eastern India. **Antimicrobial Agents Chemotherapy**, v.48, n.2, p.681-684, 2004.

PERRY, C. M.; BROGDEN, R. N. Cefuroxime axetil: a review of its antibacterial activity, pharmacokinetic properties and therapeutic efficacy. **Drugs**, Auckland, v.52, n.1, p.125-158, 1996.
PICHICHEIRO, M.E; CASEY, J.R. Bacterial eradication rates with shortened courses of 2ndande 3rd-generation cephalosporins versus 10 days of penicillin of treatment of group A streptococcal tonsillopharyngitis in adults. **Diagnostic Microbiology and Infectious Disease**, v. 59, n. 2, p. 127130, 2007.

PINTO, A.C. et al. Produtos Naturais Actualidades, Desafios e Perspectivas. **Quimica Nova**, v. 25, n. 1, p. 45-61, 2002.

PIZZO, M. A. Pattern of seed deposition and survival of seeds and seedlings of two species of Myrtaceae in the Atlantic Forest. **Revista Brasileira de Botânica**, v. 26, p.3, p.371-377. July/Sept. 2003.

PONG, A.L.; BRADLEY, J.S. Guidelines for the selection of antibacterial therapy in children. **Pediatric Clinical of North America**, v. 52, p. 860-894, 2005.

PUPO, M. T. et al. In Modern Biotechnology in Medicinal Chemistry and Industry; Taft, C. A. **Research Signpost: Kerala**, 2006.

RANG, H.P.; DALE, M.M.; RITTER, J.M. **Farmacologia**. 4. ed. Rio de Janeiro: Guanabara Koogan, 2001. 703 p.

REIS, M. S.; MARIOT, A.; STEENBOCK, W. Diversity and domestication of medicinal plants. **Simoes CMO et al. Farmacognosia: da planta ao medicamento**, v. 5, p. 45-74, 2004. ROBINEAU, L.G. **Hacia uma farmacopea caribena** / TRAMIL 7. Enda-Caribe UAG & Universidad de Antioquia, Santo Domingo, 696p. 1995.

RODRIGUES, E. A. C. et al. **Hospital Infections**: prevention and control. Sao Paulo: Sarvier, 1997.

SILVA, M.G.V. 1996. **Essential oils: Contribution to the Ocimum genus taxon and analysis by**

NMR[13]C spectrometry. 1996, 219 f. Thesis (Doctorate in Organic Chemistry) - Federal University of Ceará, Fortaleza. 1996.

SOFIATI, F. **Phytochemical study and preliminary biological activities of extracts of *Polygonum acre* (Polygonaceae) H.B.K. and *Synadenium carinatum* Boiss. (Euphorbiaceae)**. 2009. 100 f.

Dissertation (Master's Degree in Pharmaceutical Sciences) - Paulista State University, Araraquara, Sao Paulo. 2009.

SÓ, M.V.R. et al. In vitro antimicrobial activity of a propolis suspension against Enterococcus faecalis. **RFO UPF**, v. 16, n. 3, p. 277-281, 2011.

SADER, H. S. et al. Pathogen frequency and resistance patterns in Brazilian hospitals: summary of results from three years of the SENTRY Antimicrobial Surveillance Program. **Brazilian Journal of Infectious Diseases**, v. 5, n. 4, p. 200-214, 2001.

SANDES, A.R.R.; DI BLASI, G. Biodiversity and Chemical and Genetic Diversity.

Biotecnologia, n.13. p. 28-32, 2000.

SANTOS, F. O. **Biological activities of *Anacardium occidentale* (Linn)** (**Postgraduate** course in agrosilvo-pastoral systems in the semi-arid region), Federal University of Campina Grande, 2011.

SCHAPOVAL, E. E. S. et al. Evaluation of some pharmacological activities on *Eugenia uniflora* L. **Journal of Ethnopharmacology**. v. 44, p. 137-142, 1994.

SIMÔES, C. M. O. et al. **Farmacognosia: da planta ao medicamento - Produtos de origem vegetal e o desenvolvimento de medicamentos**. 5. ed. Porto Alegre/Florianópolis: UFGRS/UFSC, 2003.

SOBRAL, M.; et al. **Plantas da Floresta Atlàntica. Instituto de Pesquisas Jardim Botànico do Rio de Janeiro,** Rio de Janeiro, 2009. Available at: http://www.jbrj.gov.br/publica/livros_pdf/plantas_floresta_atlantica.pdf. Accessed on: December 1, 2015.

SOBRAL, M. **A familia Myrtaceae no Rio Grande do Sul.** Sao Leopoldo: Ed. Unisinos, 2003.

SOUZA, V.C.; LORENZI, H. **Botànica Sistemàtica - Illustrated guide to the identification of native and exotic phanerogamous families in Brazil, based on APG II.** Nova Odessa, Plantarum, 2008.

SARTORATTO, A. et al. Composition and antimicrobial activity of essential oils from aromatic plants used in Brazil. **Brazilian Journal Microbiology**, v.35, n.4, p.273-280, 2004.

TÔRRES, A.R.; OLIVEIRA, R.A.G.; DINIZ, M.F.F.M.; ARAÙJO, E.C. Study on the use of medicinal plants in hospitalized children in the city of Joao Pessoa: risks and benefits. **Revista**

Brasileira de Farmacognosia, n. 105, p.373-380 2005.

TORTORA, G.J.; FUNKE, B.R.; CASE, C.L. Microbiology, 6ª edition. **Editora Artmed, Porto Alegre**, 2003.

UEDA-NAKAMURA, T. et al. Antileishmanial activity of Eugenol-rich essential oil from *Ocimum gratissimum*. **Parasitology International**, v.55, n.2, p.99-105, 2006.

VASQUEZ, E.A. et al. Chemical and biological Studies on essential oil of Coleus amboinicus Lour. In: Annual Congress of the Society of Medicinal Plant Research. **Plantamed,** v.31:308-10,1998. VELAZQUEZ-MEZA, M. E. Methicillin-resistant Staphylococcus aureus: emergence and dissemination. **Salud Pùblica de México**, v. 47, n. 5, p. 381-7, 2005.

VENDRUSCOLO, G.S.; RATES, S.M.K.; MENTZ, L.A. Chemical and pharmacological data on the plants used as medicinals by the community of the Ponta Grossa neighborhood, Porto Alegre, Rio Grande do Sul. **Revista Brasileira de Farmacognosia** n. 15, p. 361-372, 2005.

VEIGA JÙNIOR V.F. et al. Medicinal plants: a safe cure? **Quimica Nova**, Sao Paulo, v. 28, n. 3, jun. 2005.

VOLPATO, A.M.M. **Evaluation of the antibacterial potential of *Calendula officinales* (Astearaceae) for its phytotherapeutic use.** Curitiba, 2005. 137 p. (Doctoral dissertation) Postgraduate Program in Chemistry, State University of Ponta Grossa, Curitiba, 2005. WOODFORD, N. Biological counterstrike: antibiotic resistance mechanisms of Gram-positive cocci. **Clinical Microbiology and Infection**, v. 11, n. 3, p. 2-21, 2005.

WALSH, C. **Antibiotics: Actions, Origins, Resistance**, ASM Press: Washington, 345pp. 2003. WRIGHT, G. D.; Bacterial resistance to antibiotics: Enzymatic degradation and modification. **Advanced Drug Delivery Reviews**, v. 57, n.10, p.1451-1470, 2005.

YUNES, R.A.; CALIXTO, J.B. (eds.). **Medicinal plants from the perspective of modern medicinal chemistry**. Chapecó: Argos, 2001.

CHAPTER 2

EXECUTION OF THE EXPERIMENT

Place of research

The research was carried out at the Nùcleo de Bioprospecçao e Experimentaçao Molecular Aplicada (NUBEM- INTA), located at the Instituto Superior de Teologia Aplicada, in the city of Sobral- CE.

Type of research

This is an experimental, quantitative, observational study.

Collection and identification of botanical material

Healthy leaves and flowering branches of *Eugenia uniflora* (pitanga), *Syzygium malaccense* (jambo-vermelho), *Syzygium cumini* (azeitona preta) and *Psidium guajava* (guiaba) were collected in the city of Alcântaras - Cearà in February 2014. Specimens of the plants were used to make exsiccates, where they were identified in the Prof. José de Abreu Matos herbarium at the Universidade Estadual Vale do Acaraù - UVA, in the municipality of Sobral-CE and are deposited under numbers 18852 (*Eugenia uniflora*), 18851 (*Syzigium malaccense*), 18850 (*Syzigium cumini*), 18849 (*Psidium guajava*).

The botanical material (leaves) of all the species studied from the Anacardicaeae family were collected in Ibiapina-Cearà. Specimens of the plants were used to make exsiccates, where they were identified in the Prof. José de Abreu Matos herbarium at the Universidade Estadual Vale do Acaraù - UVA, in the municipality of Sobral-CE and are deposited under numbers 20303 (*Anacardium ocidentale*), 20304 (*Mangifera indica*), 20306 (*Spondias purpurea*).

The raw material, leaves, of all the species studied from the Lamiaceae family were collected in the garden of the Sobral Phytotherapy Center (CEFISO) located in the Family Health Center in the Sumaré district. The species belonging to CEFISO are supplied by the Medicinal Plants Garden of the Phytotherapy Center of the State Health Department (NUFITO), where they are identified.

Obtaining the crude extract

After being collected, the leaves were washed and dried in the shade, without overlapping, for a period of five days (MATOS, 2007). The petioles were then removed and pulverized. 3 g of powder and 30 mL of sterile distilled water were used for each plant, using the infusion method. After cooling, they were kept in the fridge in containers protected from light. For the hydroalcoholic extract, 3 g of the powder was added to 30 mL of the solution containing sterile distilled water with 100% PA ethyl alcohol (1:1), and for the alcoholic extract, 3 g of the powder was used in 30 mL of PA alcohol. On the day of the antimicrobial test, after 72 hours of maceration, the solutions were filtered through filter paper to obtain the crude extracts.

Preparation of diluted extracts

The dilutions for the tests were made from the crude extract. Three dilutions were tested for all species: 100mg/mL, 50 mg/mL and 25 mg/mL. These were diluted in sterile containers and protected from light. Sterile distilled water was used as a negative control for the aqueous extract, a solution containing sterile distilled water with 100% ethyl alcohol PA (1:1) was used for the hydroalcoholic extract and 100% ethyl alcohol was used for the alcoholic extract.

Bacteria used

The bacteria used were: *Stretptococcus pyogenes, Enterococcus faecalis*, *Staphylococus aureus and Pseudomonas aeruginosa.* They are stored in Brain Heart Infusion (BHI) 20% glycerol in a freezer at - 20° C at the Núcleo de Bioprospecçao e Experimentaçao Molecular Aplicada (NUBEM- INTA).

Table 1 - Bacterial strains used to evaluate the antimicrobial activity of the extracts.

Species	Origin	Gram
Stretptococcus pyogenes	ATCC 6530	Gram-positive
Enterococcus faecalis	ATCC 19615	Gram-positive

Staphylococus aureus	ATCC 4083	Gram-positive
Pseudomonas aeruginosa	ATCC 154442	Gram-negative

Antibiosis tests

The antimicrobial activity of the extracts from the leaves of pitanga, black olive, jambo and guava plants was analyzed in triplicate and checked using the Mueller-Hinton agar diffusion test. After reactivation, all the strains were diluted in 0.85% sterile saline solution and adjusted according to the standard 0.5% McFarland scale (TAVEIRA, 2007). Sterile distilled water was used as a negative control for the aqueous extract, a solution containing sterile distilled water with 100% PA ethyl alcohol (1:1) was used for the hydroalcoholic extract and 100% ethyl alcohol was used for the alcoholic extract and discs containing 10 µg of Cefuroxime were used as a positive control.

Aliquots of the previously diluted bacterial cultures were taken and sown with the aid of a swab using the depletion technique on petri dishes containing Mueller-Hinton Agar.

The plates were incubated in an oven at 37°C for 24 hours. After the incubation period, the inhibition halos were measured using a ruler. Extracts that generated halos ≥ 7 mm (seven millimeters) were considered to have antimicrobial potential against the bacterial isolates (ARAUJO, 2011).

References

ARAUJO, Y.L.F.M. et al. Comparison between two techniques used in the antibacterial sensitivity test of the hydroalcoholic extract of red propolis. **Scientia Plena,** v. 7, n. 4, 2011.

MATOS, F. J. A. **Plantas Medicinais**: **guia de seleção e emprego de plantas usadas em fitoterapia no Nordeste do Brasil.** 3 ed. Fortaleza: University Press, 2007.

TAVEIRA, C.C. **Antimicrobial action of Cerrado plant extracts and isolation of active substance from *Kielmeyera coriacea*.** 2007, 74f. Dissertation (Postgraduate Diploma in Medical Sciences) - Faculty of Medicine, Brasilia. 2007.

CHAPTER 3

RESULTS ON THE LAMIACEAE FAMILY

The results of the aqueous extract of the plants tested are shown in Table 1. It did not promote the formation of halos at the concentrations tested on any of the bacteria studied.

Table 1 - Results of the antimicrobial activity of the aqueous extract of *Plectranthus amboinicus, Mentha arvensis, Ocimun gratissimum and Plectranthus barbatus* leaves at concentrations of 100 mg/mL, 50 mg/mL and 25 mg/mL.

Concentrations of aqueous extract Species	**(mg/mL)**			**CN**	**CP**
	25mg/mL	50mg/mL	100mg/mL		
Streptococcus pyogenes					
Plectranthusamboinicus	0 mm	0 mm	0 mm	0 mm	10 mm
Mentha arvensis	0 mm	0 mm	0 mm	0 mm	14 mm
Ocimum gratissimum	0 mm	0 mm	0 mm	0 mm	14 mm
Plectranthus barbatus	0 mm	0 mm	0 mm	0 mm	13 mm
Enterococcus faecalis					
Plectranthus amboinicus	0 mm	0 mm	0 mm	0 mm	15 mm
Mentha arvensis	0 mm	0 mm	0 mm	0 mm	16 mm
Ocimum gratissimum	0 mm	0 mm	0 mm	0 mm	17 mm
Plectranthus barbatus	0 mm	0 mm	0 mm	0 mm	15 mm
Staphylococcus aureus					
Plectranthus amboinicus	0 mm	0 mm	0 mm	0 mm	23 mm
Mentha arvensis	0 mm	0 mm	0 mm	0 mm	20 mm

Ocimum gratissimum	0 mm	0 mm	0 mm	0 mm	22 mm
Plectranthus barbatus	0 mm	0 mm	0 mm	0 mm	22 mm
Pseudomonas aeruginosa					
Plectranthus amboinicus	0 mm	0 mm	0 mm	0 mm	12 mm
Mentha arvensis	0 mm	0 mm	0 mm	0 mm	11 mm
Ocimum gratissimum	0 mm	0 mm	0 mm	0 mm	13 mm
Plectranthus barbatus	0 mm	0 mm	0 mm	0 mm	13 mm

CN: Negative control; CP: Positive control (Cefuroxime)

According to Silva et al. (2002), of the estimated 40,000 to 55,000 plant species in the Brazilian flora, little is known about their chemical composition. In addition, there are a number of factors that can interfere with their composition and action, such as the time, form and place where the plants are collected, the type of soil, altitude, solar radiation, climatic changes, among others (CECHINEL; YUNES, 1998; ALVES; TRESMONDI; LONGUI 2008; LÓPEZ, 2010).

The work published by Haida et al. (2007), using the aqueous extract of *Ocimum gratissimum*, observed no antimicrobial activity against *Staphylococcus aureus and Pseudomonas aeuriginosa* bacteria at concentrations of 500 mg/ML and diluted to 10% (100 mg/mL), 5% (50 mg/mL), 2.5% (25 mg/mL), 1% (10 mg/mL) and 0.1% (1 mg/mL). The methodology used by Haida et al. (2007) was similar to the present study when they used similar concentrations and used Mueller-Hinton agar, differing in terms of extraction: they used solvents of increasing polarity (n-hexane, chloroform, acetone, and 80% ethanol, aqueous (infusion), and the antimicrobial action tests were carried out using the diffusion method on filter paper disks by adapting the Kirby-Bauer method.

In a study carried out by Castillo and Gonzalez (1999), the aqueous and fluid extracts of *Plectranthus amboinicus* showed a broad spectrum of antimicrobial activity for gram-positive and gram-negative bacteria, which was not the case in this study. These differences can be explained by factors such as the area, collection period, geographical location. In a study carried out by Costa (2002), she found no antimicrobial activity in the aqueous extract of *Plectranthus barbatus*, which is in agreement with the present study.

In the present study, the hydroalcoholic extract of *Plectranthus barbatus* showed antimicrobial activity against *S, aureus* at three concentrations: 100 mg/mL, 50 mg/mL, 25 mg/mL, with a halo formation of 10 mm, 9 mm and 8 mm, respectively, and against *P, aueruginosa* at a concentration of 100 mg/mL with a halo of 8 mm. The species *Plectranthus amboinicus, Mentha arvensis, Ocimum gratissimum* showed antimicrobial activity at a concentration of 100 mg/mL against the microorganism *S, aureus* with halos of 7 mm, 9 mm and 7 mm, in that order. No plant showed antimicrobial activity against the microorganisms *Enterococcus faecalis* and *Streptococcus pyogenes.* Table 2 shows the results of the hydroalcoholic extract.

Table 2 - Results of the antimicrobial activity of the hydroalcoholic extract of *Plectranthus amboinicus, Mentha arvensis, Ocimun gratissimum and Plectranthus barbatus* leaves at concentrations of 100 mg/ml, 50 mg/ml and 25 mg/ml.

Concentrations of hydroalcoholic extract					
Species	**(mg/mL)**			**CN**	**CP**
	25mg/mL	50mg/mL	100mg/mL		
Streptococcus pyogenes					
Plectranthusamboinicus	0 mm	0 mm	0 mm	0 mm	11 mm
Mentha arvensis	0 mm	0 mm	0 mm	0 mm	13 mm
Ocimum gratissimum	0 mm	0 mm	0 mm	0 mm	11 mm
Plectranthus barbatus	0 mm	0 mm	0 mm	0 mm	11 mm
Enterococcus faecalis					
Plectranthusamboinicus	0 mm	0 mm	0 mm	0 mm	10 mm
Mentha arvensis	0 mm	0 mm	0 mm	0 mm	12 mm

Ocimum gratissimum	0 mm	0 mm	0 mm	0 mm	11 mm
Plectranthus barbatus	0 mm	0 mm	0 mm	0 mm	12 mm
Staphylococcus aureus					
Plectranthusamboinicus	0 mm	0 mm	7 mm	0 mm	22 mm
Mentha arvensis	0 mm	0 mm	9 mm	0 mm	21 mm
Ocimum gratissimum	0 mm	0 mm	7 mm	0 mm	22 mm
Plectranthus barbatus	8 mm	9 mm	10 mm	0 mm	22 mm
Pseudomonas aeruginosa					
Plectranthusamboinicus	0 mm	0 mm	0 mm	0 mm	11 mm
Mentha arvensis	0 mm	0 mm	0 mm	0 mm	10 mm
Ocimum gratissimum	0 mm	0 mm	0 mm	0 mm	11 mm
Plectranthus barbatus	0 mm	0 mm	8 mm	0 mm	11 mm

CN: Negative Control; CP: Positive Control (Cefuroxime)

In 1970, Bhakuni et al. screened 295 plants from India, including *P. barbatus*, for antimicrobial activity against gram-positive, gram-negative and alcohol-resistant bacteria. A hydroalcoholic extract of the whole plant was tested, except for the roots, which showed no antimicrobial action.

In agreement with this study, Gurgel et al. (2007) found that the hydroalcoholic extract of *P. amboinicus* leaves showed antimicrobial action against gram-positive bacteria, represented in the study by *S. aures*, forming halos of between 16 mm and 18 mm. *P. amboinicus* is chemically composed of essential oils (thymol, carvacrol), flavanoids and tannins, all of which are known to have antimicrobial properties. In the same study, it was observed that there was no antimicrobial activity of the hydroalcoholic extract against gram-negative bacteria, represented by two strains of *Salmonella*. The method used by these authors was similar to the one used in the present study. They used the maceration technique to obtain the extracts, but they were filtered in a rotaevaporator and the antimicrobial activity was determined using the solid medium diffusion technique, where perforations were made in the culture medium containing Mueller-Hinton agar.

Lund et al. (2012) carried out a study using a hydroalcoholic extract of the aerial parts, leaves and stems of *Mentha arvensis* against gram-positive bacteria and fungi, *Streptococcus mutans, Streptococcus sobrinus and Candida albicans*. The extract was obtained by maceration at room temperature for nine hours and the antimicrobial activity was measured using agar diffusion tests using the disk diffusion method. He also found that *M. arvensis* did not show antimicrobial activity for these pathogens, but did show antifungal activity. However, in the present study, it was observed that this same species showed antimicrobial action against S. aureus at a concentration of 100 mg/mL.

According to Franco et al. (2007) the essential oil of *O. gratissimum* shows growth inhibitory activity against *S. aureus*, *E. coli* and *S. choleraeseus* bacteria. Hommerding et al. (2014) studied the antimicrobial activity of the essential oil of *O. gratissimum* and its pure oil showed action against gram-positive and gram-negative bacterial strains. The diversity of biological activities of this plant species is due to the existence of a wide variety of compounds present in it, one of the main ones cited in its essential oil is eugenol, which is one of the compounds that explains the antimicrobial activity of the essential oil.

Table 3 shows the results of the alcoholic extract, which showed antimicrobial activity for three species of bacteria, with only *Enterococcus faecalis* showing no antimicrobial activity. For *Streptococcus pyogenes*, a halo was formed on two plants at a concentration of 100 mg/mL*, Mentha arvensis and Plectranthus barbatus*, with a size of 9 mm and 7 mm, respectively. In the case of *Staphylococcus aureus, the species Plectranthus amboinicus, Mentha arvensis, Plectranthus barbatus* showed halo formation of 9 mm, 8 mm and 10 mm, with only *Ocimun gratissimum* showing no antimicrobial activity. Finally, *Plectranthus barbatus* showed antimicrobial activity with a halo formation of 9 mm for *Pseudomonas aeuroginosa.*

Table 3 - Results of the antimicrobial activity of the alcoholic extract of *Plectranthus* *amboinicus, Mentha arvensis, Ocimun gratissimum and Plectranthus barbatus* leaves at concentrations of 100 mg/mL, 50 mg/mL and 25 mg/mL.

Concentrations of the alcoholic extract					
Species	**(mg/mL)**			**CN**	**CP**
	25mg/mL	50mg/mL	100mg/mL		
Streptococcus pyogenes					
Plectranthus amboinicus	0 mm	0 mm	0 mm	0 mm	10 mm
Mentha arvensis	0 mm	0 mm	9 mm	0 mm	10 mm
Ocimum gratissimum	0 mm	0 mm	0 mm	0 mm	13 mm
Plectranthus barbatus	0 mm	0 mm	7 mm	0 mm	11 mm
Enterococcus faecalis					
Plectranthus amboinicus	0 mm	0 mm	0 mm	0 mm	11 mm
Mentha arvensis	0 mm	0 mm	0 mm	0 mm	11 mm
Ocimum gratissimum	0 mm	0 mm	0 mm	0 mm	13 mm
Plectranthus barbatus	0 mm	0 mm	0 mm	0 mm	11 mm
Staphylococcus aureus					
Plectranthus amboinicus	0 mm	0 mm	9 mm	0 mm	22 mm
Mentha arvensis	0 mm	0 mm	8 mm	0 mm	21 mm
Ocimum gratissimum	0 mm	0 mm	0 mm	0 mm	22 mm
Plectranthus barbatus	0 mm	0 mm	10 mm	0 mm	20 mm
Pseudomonas aeruginosa					
Plectranthus amboinicus	0 mm	0 mm	0 mm	0 mm	11 mm
Mentha arvensis	0 mm	0 mm	0 mm	0 mm	10 mm
Ocimum gratissimum	0 mm	0 mm	0 mm	0 mm	11 mm

Plectranthus barbatus	0 mm	0 mm	9 mm	0 mm	10 mm

NC: Negative Control; PC: Positive Control (Cefuroxime)

In a study carried out by Costa (2002), it was found that the alcoholic extract of the root, stem and leaves of *P. barbatus* were active against gram-positive and alcohol-resistant bacteria, but were not active against gram-negative bacteria such as *Monilia sitophita* and *Candida albicans*. The extracts used by this author were obtained by macerating for eight days under agitation and at a temperature of 28 to 30 °C. This confirms the results of this research in parts, where it showed activity against *Streptococcus pyogenes, Staphylococcus aureus* (gram-positive), and differed with regard to gram-negative bacteria.

Nogueira (2004) observed that the crude extract of *P. amboinicus* leaves showed antimicrobial activity in patients with otitis externa. Oliveira (2006) observed that the essential oil of *P. amboinicus*, when combined with flavones, aromatic acids and tannins, showed antimicrobial activity against strains *of Pneumococcus and Staphylococcus.*

According to Haida et al. (2007), the alcoholic extract of *Ocimum gratissimum* showed no antimicrobial activity, which is consistent with this study. Matias et al. (2010) showed that the methanolic and hexanolic extracts of *Ocimum gratissimum* had significant antimicrobial activity. According to Matos (2002), the essential oil has a local antiseptic action against *S. aureus*.

In a study carried out by Pereira et al. (2004), they found antimicrobial action of the essential oil of *O. gratissimum* against 55 out of 79 strains of *E. coli*, and with even less efficacy against *P. aeruginosa*.

According to Johnson et al. (2011), the ethanolic, methanolic and hexanolic extracts, among other solvents used, of leaves and inter-nodal segments derived from the callus of the species *Mentha arvensis* showed antimicrobial activity against the bacteria *Salmonella typhy, Streptococus pyogenes, Proteus vulgaris and Bacillus subtilis.* This confirms the results of this study, in which the ethanolic extract of *Mentha arvensis* leaves showed antimicrobial activity against *S. pyogenes.*

The use of plants and their constituents with known antimicrobial action is of great therapeutic importance in the treatment of infections (NASCIMENTO et al., 2000). According to Fernandes et al. (2011), studies of plant extracts and oils found that the antimicrobial activity of plant species is due to products of secondary metabolism, which in their pure form show action. The lack of conformity in research and in the literature may be associated with the presence and quantity of each metabolite present in the samples, the method used and the period of the year when the samples were taken.

In addition, everything that surrounds the plants can interfere with the chemical composition of the species, such as temperature, age of the plant, location and water availability.

Valgas (2002) reports that in the literature there are many and varied studies on the quantity and concentrations of extracts and substances used in tests to identify antimicrobial activity, and it is necessary for researchers to combine similar techniques and concentrations of extracts in order to help and improve interpretation and relate the results achieved.

References

ALVES, E.S.; TRESMONDI, F.; LONGUI, E.L. Structural analysis of *Eugenia uniflora* L. (Myrtaceae) leaves collected in rural and urban environments, SP, Brazil. **Acta Botànica Brasilica**, v. 95, n. 1, p 367-373, 2008.

BHAKUNI, D. S. et al. Sreening of Indian plants for biological activity: part III. **Indian Journal of Experimantal Biology**, v. 9, p. 91-102, 1970.

CECHINEL FILHO, V.; YUNES, R. A. Strategies for obtaining pharmacologically active compounds from medicinal plants: concepts on structural modification to optimize activity. **Quimica Nova**, v. 21, n. 1, p. 99-105, 1998.

CASTILLO, R. A. M.; GONZALEZ, V. P. *Plecthranthus amboinicus* (lour.) spreng. **Revista Cubana Plantas Medicinales,** v. 3, n. 3, p. 110-115. 1999.

COSTA, M.C.C.D. **Pharmacological aspects of Plectranthus barbatus Andr. (Lamiaceae): antimicrobial, cytotoxic and antitumor activities**. 2002.124p.

Thesis (Doctorate in Biological Sciences)-Federal University of Pernambuco, Recife, 2002. COSTA,

M.C.C.D. **Aspectos farmacológicos de Plectranthus barbatus Andr. (Lamiaceae): atividades antimicrobiana, citotóxica e antitumoral**. 2002.124p. Thesis (Doctorate in Biological Sciences)- Federal University of Pernambuco, Recife, 2002.

FERNANDES, A.P. et al. Effect of the hydroalcoholic extract of *Pyrostegia venusta* on mutagenesis *"in vivo", and* antimicrobial evaluation, and interference in cell growth and differentiation *"in vitro"*. **Revista Medicina Minas Gerais**. v. 21, n. 3, p. 272-279, 2011.

FRANCO, A.L.P. Evaluation of the chemical composition and antibacterial activity of the essential oils of *Aloysia gratissima* (Gillies & Hook) *Tronc.(*lavender), Ocimum gratissimum L.(wild lavender) and *Curcuma longa* L.(safflower). **Revista Eletrônica de Farmàcia**, v. 4, n. 2, p. 208-220, 2007.

GURGEL, A.P.A.D. **The importance of *Plectranthus amboinicus* (Lour.) Spreng as a therapeutic alternative - Experimental methods.** Pernambuco. 2007. 89 p. Dissertation (Master's Degree in Pharmaceutical Sciences). Federal University of Pernambuco. Recife, 2007. HOMMERDING, L.R. et al. *In vitro* antimicrobial activity of the essential oil of *Ocimum gratissimum* L. (Alfavaca-cravo). **Revista de Biologia e Saùde da Unisp**, v. 7, n.1, p. 1-9, 2014. HAIDA, K. S. et al. In vitro evaluation of the antimicrobial activity of eight medicinal plant species**. Arquivos de Ciències da Saùde da Unipar, Umuarama,** v. 11, n. 3, p. 185-192, 2007. JOHNSON, M. et al. Antibacterial activity of leaves and inter-nodal callus extracts of *Mentha arvensis* L. **Asian Pacific journal of tropical medicine**, v. 4, n. 3, p. 196-200, 2011.

LÓPEZ, P. V. A. **Bioprospecting of extracts from *Croton urucurana* Baillon and its endophytic fungi**, 138f Dissertation (Master's Degree in Microbiology) - Federal University of Paranà, UFPR, Curitiba. 2010.

LUND, R.G. et al. In vitro study on the antimicrobial effect of hydroalcoholic extracts from *Mentha arvensis* L.(Lamiaceae) against oral pathogens-doi: 10.4025/actascibiolsci. v34i4. 8959. **Acta Scientiarum. Biological Sciences**, v. 34, n. 4, p. 437-442, 2012.

NOGUEIRA J.C.R. ***In vitro* antimicrobial activity of plant products on microorganisms isolated from patients with acute otitis externa**. Joao Pessoa. 2004. 180 p. Dissertation (Master's Degree in Synthetic and Bioactive Products). Federal University of Paraiba, 2004.

NASCIMENTO, G.G.F. et al. Antibacterial activity of plant extracts and phytochemicals on antibiotic-resistant bacteria. **Brazilian Journal of Microbiology**, v.31, n.4, p.247-56, 2000.

OLIVEIRA R.A.G. **Medicinal plants traditionally used in Dermatology: Evaluation of the biological activity of their extracts, essential oils and their associations**. Joao Pessoa. 2006. 180 p. Thesis (Doctorate in synthetic and bioactive products). Federal University of Paraiba, 2006.

PEREIRA, R.S. et al. Antibacterial activity of essential oils in strains isolated from urinary infection. **Revista Saùde Pùblica**, v. 38, n. 2 p. 326-328. Sao Paulo, 2004.

SILVA, R. L. et al. Effect of the aqueous extract of *Hyptispectinata* on hepatocyte proliferation after partial hepatectomy. **Acta Cirùrgica Brasileira**, v.17. Supplement 3, p.101-105, 2002.

VALGAS, C. **Evaluation of screening methods for determining the antimicrobial activity of natural products**. 2002. 91f. Dissertation (Master's Degree in Pharmacy) - Universidade Federal Santa Catarina, Florianópolis, Santa Catarina, 2002.

CHAPTER 4

RESULTS ON THE MYRTACEAE FAMILY

When analyzing the results of the aqueous extract, *Eugenia uniflora, Syzygium cumini and Psidium guajava* showed antimicrobial activity against *Staphylococcus aureus* with a halo of 12 mm, 8 mm and 7 mm at a concentration of 100 mg/m, respectively. Table 1 summarizes the results of the antibiosis tests with the aqueous extract.

Table 1- Antimicrobial activity of aqueous extracts of *Eugenia uniflora*, *Syzygium malaccense*, *Syzygium cumini* and *Psidium guajava* leaves at concentrations of 25 mg/mL, 50 mg/mL and 100 mg/mL.

Concentrations of the aqueous extract					
Extract	**(mg/mL)**			**CN**	**CP**
Sthaphylococcus aureus					
	25 mg/mL	50mg/mL	100mg/Ml		
Eugenia uniflora	10 mm	12 mm	12 mm	0 mm	23 mm
Syzygium malaccense	0 mm	0 mm	0 mm	0 mm	22 mm
Syzygium cumini	0 mm	0 mm	8 mm	0 mm	25 mm
Psidium guajava	0 mm	0 mm	7 mm	0 mm	24 mm
Streptococcus pyogenes					
Eugenia uniflora	0 mm	0 mm	0 mm	0 mm	14 mm
Syzygium malaccense	0 mm	0 mm	0 mm	0 mm	17 mm
Syzygium cumini	0 mm	0 mm	0 mm	0 mm	15 mm
Psidium guajava	0 mm	0 mm	0 mm	0 mm	14 mm
Enterecoccus faecalis					
Eugenia uniflora	0 mm	0 mm	0 mm	0 mm	13 mm

Syzygium malaccense	0 mm	0 mm	0 mm	0 mm	12 mm
Syzygium cumini	0 mm	0 mm	0 mm	0 mm	14 mm
Psidium guajava	0 mm	0 mm	0 mm	0 mm	18 mm
Pseudomonas aeruginosa					
Eugenia uniflora	0 mm	0 mm	0 mm	0 mm	12 mm
Syzygium malaccense	0 mm	0 mm	0 mm	0 mm	12 mm
Syzygium cumini	0 mm	0 mm	0 mm	0 mm	13 mm
Psidium guajava	0 mm	0 mm	0 mm	0 mm	11 mm

* Positive control (PC): Cefuroxime. Negative control (NC): Sterile distilled water.

When trying to compare studies with plants, it is undeniably difficult to evaluate the results as a result of the action of various variables, ranging from aspects relating to the soil and climate that influence the chemical composition, as well as the stage of development of the plant, form of collection, part of the plant to be studied, form of preparation of the material for study, to the protocols that will be followed in the experiments (AURICCHIO; BACCHI, 2003).

For Alves, Tresmondi and Longui (2008) the fact that the plant was collected in a rural or urban area can influence the antimicrobial activity of the extract. Other factors such as altitude, soil type, solar radiation (CECHINEL; YUNES, 1998; LÓPEZ, 2010), extreme changes as well as stressful situations such as fires and floods can also interfere with the production of chemical compounds (BEZERRA, 2013).

When comparing the results of the inhibitory responses (halo diameter) obtained in this study with the literature, it can be seen that the plants researched show different responses in relation to antimicrobial activity. For Alves et al. (2000), Holetz et al. (2002), Gonçalves, Alves and Meneses (2005), Lopes (2006) the formation of inhibition halos under the microorganisms tested is due to a synergistic effect of all their constituents, phytochemical compounds: tannin, phenols, flavonoids and alkaloids.

According to Didry et al. (1998), *Psidium guajava* leaf extract was able to inhibit the bacterial growth of *Sthaphylococcus aureus* (ATCC 25923), and he also observed antibacterial activity against *Streptococcus oralis* and *Streptococcus mitis*. Similar results were found in this study, as *Psidium guajava* also inhibited *Sthaphylococcus aureus*.

Testing aqueous extracts of *Psidium guajava*, Menezes et al. (2004) found positive results for antimicrobial activity, with inhibition halos averaging 25 mm for *Sthaphylococcus aureus*. These results confirm those obtained in this study, but there is a divergence in the size of the halos which may be explained by the methodology used to obtain the extracts or the concentration used.

According to Oliveira (2007), the infusion of the fresh leaf of *Eugenia uniflora* was the herbal medicine that best exerted antibacterial activity against cariogenic bacteria. This agrees with the results of this study, which indicated the antimicrobial activity of pitanga in aqueous extract against *Staphylococcus aureus*.

Pessini et al. (2003) in a study evaluated 13 plants used in folk medicine to treat infectious diseases. Among these plants, *Eugenia uniflora* inhibited the development of some strains of bacteria and fungi, which reinforces the results of this study that showed inhibition of *Staphylococcus aureus* and reaffirms the popular belief in the use of this plant in the treatment of various infectious diseases.

Chattopahyay, Sinha and Vaid (1998) carried out studies with aqueous and alcoholic extracts (95%) of *Syzygium cumini*, showing inhibition at varying concentrations against Gram-positive bacteria (*Staphylococcus aureus and Bacillus subtilis*) and most Gram-negative bacteria (*Escherichia coli, Salmonella typhi, Morganella morgani and Pseudomonas aeruginosa*). This study reaffirms the antimicrobial activity against *Staphylococcus aureus*, but differs in relation to the inhibition of *Pseudomonas aeruginosa*, as it did not show a halo of inhibition with the aqueous extract.

Locher and colleagues (1995) tested the antibacterial action of the aqueous extract of *Syzygium malaccense* leaves using the solid medium diffusion method and observed a inhibition of the growth of *Streptococcus pyogenes* and *Staphylococcus aureus* strains at a concentration of 100 µg/mL. This was different to what was observed in this study, as there was no inhibition of any of the bacteria studied.

According to Rodrigues et al. (2003), the aqueous extract of *Syzygium malaccense* leaves showed greater sensitivity against Gram-positive bacteria *Staphylococcus aureus.* This disagrees with the present study, which showed no activity in the aqueous extract at all the concentrations tested.

In the data analysis of the hydroalcoholic extracts, *Syzygium cumini* showed antimicrobial activity with a halo of 7 mm at a concentration of 100 mg/ml on *Pseudomonas aeruginosa.* However, the antimicrobial activity of the plants studied on *Streptococcus pyogenes and Enterecoccus faecalis* was not observed. Table 2 summarizes the results of the antibiosis tests with the hydroalcoholic extract.

Table 2 - Antimicrobial activity of hydroalcoholic extracts of *Eugenia uniflora*, *Syzygium malaccense*, *Syzygium cumini* and *Psidium guajava* leaves at concentrations of 25 mg/mL, 50 mg/mL and 100 mg/mL.

Concentrations of hydroalcoholic extract					
Extract	**(mg/mL)**			**CN**	**CP**
Staphylococcus aureus					
	25 mg/mL	50mg/mL	100mg/mL		
Eugenia uniflora	12 mm	12 mm	14 mm	0 mm	20 mm
Syzygium malaccense	0 mm	0 mm	7 mm	0 mm	16 mm
Syzygium cumini	7 mm	8 mm	10 mm	0 mm	20 mm
Psidium guajava	7 mm	8 mm	9 mm	0 mm	22 mm
Streptococcus pyogenes					
Eugenia uniflora	0 mm	0 mm	0 mm	0 mm	13 mm
Syzygium malaccense	0 mm	0 mm	0 mm	0 mm	13 mm
Syzygium cumini	0 mm	0 mm	0 mm	0 mm	14 mm

Psidium guajava	0 mm	0 mm	0 mm	0 mm	14 mm
Enterecoccus faecalis					
Eugenia uniflora	0 mm	0 mm	0 mm	0 mm	12 mm
Syzygium malaccense	0 mm	0 mm	0 mm	0 mm	13 mm
Syzygium cumini	0 mm	0 mm	0 mm	0 mm	10 mm
Psidium guajava	0 mm	0 mm	0 mm	0 mm	10 mm
Pseudomonas aeruginosa					
Eugenia uniflora	0 mm	0 mm	0 mm	0 mm	11 mm
Syzygium malaccense	0 mm	0 mm	0 mm	0 mm	11 mm
Syzygium cumini	0 mm	0 mm	7 mm	0 mm	11 mm
Psidium guajava	0 mm	0 mm	0 mm	0 mm	11 mm

* Positive control (PC): Cefuroxime. Negative control (NC): Solution containing sterile distilled water with 100% PA ethyl alcohol (1:1).

The differences in activity against gram-positive and gram-negative bacteria may derive from the constitution of the bacterial cell wall and the constituents of the plant extract, particularly the tannin group. According to some authors (KHAN et al., 2001; SRINAVASAN et al., 2001; CIMANGA et al., 2002), there is a relationship between tannin content and activity against Gram-positive bacteria, which have a more rigid cell structure, a chemically less complex cell wall and a lower lipid content than Gram-negative bacteria.

Holetz et al. (2002) observed antimicrobial activity of the hydroalcoholic extract of *Psidium guajava* leaves against *Candida albicans*, *Staphylococcus aureus* and *Escherichia coli.*

Sanches et al. (2005) also obtained an inhibitory effect using the hydroalcoholic and aqueous extracts of *Psidium guajava* against standard strains *of Staphylococcus aureus.* These results are similar to those obtained in this study, which showed halos of inhibition in both the aqueous and hydroalcoholic extracts.

Gonçalves et al. (2008) observed that *Streptococcus pyogenes, Proteus mirabilis* and *Staphylococcus aureus* were sensitive to *Psidium guajava* extract, *while* no action was found against *Escherichia coli and Pseudomonas aeruginosa*. The divergence of results for some strains can be explained by the methodology used to obtain the extracts, which suggests that they have different biosynthetic and chemical classes.

The extract that showed high antimicrobial activity against *staphylococcus aureus* was that of *Eugenia uniflora*, which is cited in the literature as also having activity against *Streptococcus, Escherichia coli, and Bacillus cereus* and resistance *to Pseudomonas aeruginosa* (ALMEIDA ALVES et al., 2000), this resistance was also demonstrated in the results of this study, however, no antimicrobial activity was observed against *Streptococcus pyogenes*.

Analyzing the antimicrobial activity of *Eugenia uniflora*, Auricchio (2007) observed that the hydroalcoholic extract of the leaves of this species exhibited *a* Minimum Inhibitory Concentration of 100 µg/mL against *Salmonella choleraesuis* and 80 µg/mL against *Sthaphylococcus aureus*. The above data corroborates the assertion that pitangueira extract has antimicrobial activity against *Sthaphylococcus aureus*.

Schapoval et al. (1994), in studies of preparations used in folk medicine, found no antimicrobial activity in infusions and decoctions of *Eugenia uniflora* leaves, perhaps because these preparations were diluted below the minimum inhibitory concentration.

The antimicrobial action of *Eugenia uniflora* against these strains was probably due to the presence of the phytoconstituents: tannins, phenols, flavonoids, alkaloids and saponins. According to Costa et al. (2005), these substances result from the secondary metabolism of plants and their function is to defend against predators or attract pollinators, but they can also have other biological activities.

It is interesting to note that the extract of *Syzygium cumini* shows antimicrobial activity against Gram-positive (*Staphylococcus aureus*) and Gram-negative (*Pseudomonas aeruginosa*) bacteria, which are different in terms of the structure of their membrane (SILVA, 2010) usually Gram-negative bacteria show resistance to antibiotics due to the decrease in permeability that occurs in their outer membrane (MOREIRA;

FREIRE, 2011), the results suggest that flavonoids, saponins and alkaloids may be contributing to the antimicrobial activity observed against Gram-negative strains and especially against Gram-positive strains.

Louguercio et al. (2005) used the disk diffusion method to evaluate the antimicrobial activity of the hydroalcoholic extract of jambolao leaves (*Syzygium cumini)* and found that it was active against both Gram-positive and Gram-negative bacteria. These data are in line with the findings of the present study, where the inhibition halos were greater against gram-positive bacteria.

Lima (2012) tested the hydroalcoholic extract of *Syzygium cumini* leaves in the antimicrobial evaluation using the agar-well diffusion method and observed antibacterial activity against *Staphylococcus aureus, Escherichia* and *Pseudomonas aeruginosa.*

The antimicrobial activity of *Syzygium malaccense* extract may be related to the presence of terpenoid compounds. The mechanism of action of terpenoids is not yet fully understood, but it is believed to involve the disruption of the cell membrane by lipophilic compounds (COWAN, 1999). These data are in line with the work of Abu-Shanab et al. (2004), who observed that the resistance of Gram-negative bacteria to plant extracts, such as *Escherichia coli* and *Pseudomonas aeruginosa*, is not uncommon, because this class of bacteria can offer a barrier to permeability due to the chemical composition of the cell wall.

According to Rodrigues (2009), of the eight samples of *Staphylococcus aureus* studied, seven with a multidrug resistance phenotype were inhibited with just 0.44 ± 0.28 mg/mL of the aqueous extract of *Syzygium malaccense*. This extract showed greater sensitivity against Gram-positive bacteria, especially *Staphylococcus aureus.* Similar results were found in this study, where *Staphylococcus aureus* was the only strain that showed sensitivity to the *Syzygium malaccense* extract.

In the data analysis of the alcoholic extracts, *Psidium guajava and Syzygium cumini* showed antimicrobial activity against the four strains studied. On *Strepctococcus pyogenes*, halos of 10 mm and 11 mm were observed respectively at a concentration of 100 mg/mL each.

Halos of 8 mm and 11 mm were observed against *Enterococcus faecalis*, respectively, at a concentration of 100 mg/mL. They also showed inhibition halos against *Pseudomonas aeruginosa* of 8 mm and 9 mm respectively, at a concentration of 100 mg/mL.

And on *Staphylococcus aureus*, *Eugenia uniflora, Psidium guajava* and *Syzygium cumini* showed activity with halos of 11 mm, 8 mm and 10 mm respectively at a concentration of 100 mg/ml each. The alcoholic extract of *Syzygium malaccense* showed no antimicrobial activity against the strains studied. Table 3 summarizes the results of the antibiosis tests with the alcoholic extract.

Table 3 - Antimicrobial activity of alcoholic extracts of *Eugenia uniflora*, *Syzygium malaccense, Syzygium cumini* and *Psidium guajava* leaves at concentrations of 25 mg/mL, 50 mg/mL and 100 mg/mL.

Extract	**Concentrations of alcoholic extract (mg/mL)**			**CN**	**CP**
Staphylococcus aureus					
	25 mg/mL	50mg/mL	100mg/mL		
Eugenia uniflora	8 mm	10 mm	11mm	0 mm	23 mm
Syzygium malaccense	0 mm	0 mm	0 mm	0 mm	18 mm
Syzygium cumini	7 mm	8 mm	10 mm	0 mm	24 mm
Psidium guajava	0 mm	7 mm	8 mm	0 mm	18 mm
Streptococcus pyogenes					
Eugenia uniflora	0 mm	0 mm	0 mm	0 mm	13 mm
Syzygium malaccense	0 mm	0 mm	0 mm	0mm	14 mm
Syzygium cumini	7 mm	9 mm	11 mm	0 mm	13 mm
Psidium guajava	0 mm	7 mm	10 mm	0 mm	24 mm

Enterecoccus faecalis					
Eugenia uniflora	0 mm	0 mm	0 mm	0 mm	12 mm
Syzygium malaccense	0 mm	0 mm	0 mm	0 mm	10 mm
Syzygium cumini	8 mm	10 mm	11 mm	0 mm	13 mm
Psidium guajava	0 mm	7 mm	8 mm	0 mm	13 mm
Pseudomonas aeruginosa					
Eugenia uniflora	0 mm	0 mm	0 mm	0 mm	11 mm
Syzygium malaccense	0 mm	0 mm	0 mm	0 mm	12 mm
Syzygium cumini	7 mm	8 mm	9 mm	0 mm	12 mm
Psidium guajava	0 mm	7 mm	8 mm	0 mm	10 mm

* Positive control (PC): Cefuroxime. Negative control (NC): ethyl alcohol 100% PA.

Rodrigues (2009) analyzed the antimicrobial activity of *Syzygium malaccense* extracts and found that the ethyl acetate extract was active against both Gram-negative bacteria, *Pseudomonas aeruginosa*, and Gram-positive bacteria, *Sthaphylococcus aureus*, with significantly better efficacy against *Sthaphylococcus aureus* ($p<0.5$).

The alcoholic extract of *Syzygium malaccense* in this study showed no antimicrobial activity against the strains studied. Vale and Orlanda (2011) point out that the absence of antimicrobial activity may be due to the presence of low concentrations of the phytoconstituents, which ends up reflecting the plant's low activity, as the respective strains proved resistant to the extracts at all concentrations.

Gonçalves et al. (2008), in their *in vitro* studies with microorganisms extracted from shrimp, observed that the best inhibitory effect was with the methanolic extract of *Psidium guajava* leaves against *Staphylococcus aureus*. The inhibition halos ranged from 8.25 to 9.25 mm at an estimated concentration of 96.87 µg/mL.

Pereira et al. (2010) tested and proved the antimicrobial action of *Psidium guajava* glycol extracts against standard strains *of Staphylococcus aureus*.

Betoni et al. (2006) used ethanolic extracts of guava leaves at a concentration of 131.75 μg/ml against isolates of *Staphylococcus aureus*, and managed to reduce the MIC from 52 to 13 μg/ml, by associating this extract with antibiotics (tetracycline, chloramphenicol, erythromycin, vancomycin, oxacillin, cephalothin, ampicillin, cefoxitin, cotrimoxazole) and obtained an efficacy of 47.68%.

The results obtained in the current study are consistent with the history of the antimicrobial activity of the plant species studied, in terms of microbicidal action and the demonstration that isolates of the strains studied were sensitive to the alcoholic extract used. It is important to note that the *Psidium guajava* extract showed antimicrobial activity against Gram-positive (*Staphylococcus aureus, Streptococcus pyogenes and Enterococcus faecalis*) and Gram-negative (*Pseudomonas aeruginosa*) *strains*, which differ in terms of their membrane structure. However, it is worth noting that the analysis of antimicrobial activity in the various studies cited was carried out using different methodologies and concentrations, which makes a reliable comparative analysis difficult.

Bouzada (2007), when evaluating the antimicrobial action of 41 plants used in Brazilian folk medicine, including *Eugenia uniflora*, did not observe the formation of inhibition halos, although the authors identified the presence of phytoconstituents.

Testing the antimicrobial activity *of Eugenia uniflora*, Auricchio (2007) observed that *Staphylococcus aureus, Salmonella choleraesuis, Pseudomonas aeruginosa* and *Candida albicans* were sensitive to the action of the extract. The highest levels of antimicrobial action were observed against *Sthaphylococcus aureus* and *Salmonella choleraesuis.*

When comparing the results of the inhibitory responses (halo diameter) obtained in this study with the literature, it can be seen that *Eugenia uniflora* shows different responses with regard to antimicrobial activity, with the alcoholic extract in this study showing halos only against *Sthaphylococcus aureus.*

Turatti et al (2006), analyzing the antimicrobial action of *Syzygium cumini,* considered the following concentrations determined for the evaluation of antimicrobial activity:

78.12 to 2.500.0 μg/mL, using the microorganisms *Sthaphylococcus aureus, Sthaphylococcus epidermidis, Escherichia coli, Pseudomonas aeruginosa, Candida krusei, Candida albicans* and *Candida parapsilosis*, the most active results when determining the minimum inhibitory concentration he obtained were for the microorganisms *Sthaphylococcus aureus, Sthaphylococus epidermidis* and *Candida parapsilosis*, which proved to be the most sensitive to the extract.

When analyzing the antimicrobial activity of essential oils extracted from the leaves of *Syzygium cumini* and other species of the same genus using the agar diffusion test, Shafi et al. (2002) showed antimicrobial activity values for the halo of inhibition of around 20 mm. For this study, the authors used the following microorganisms: *Sthaphylococcus aureus, Escherichia coli, Pseudomonas aeruginosa* and *Salmonella typhimurum*.

The results obtained in this study are in line with data in the literature, which indicate that *Syzygium cumini* extracts are less effective against Gram-negative bacteria, such as *Pseudomonas aeruginosa*, even though they were sensitive to the alcoholic extract of *Syzygium cumini, they* showed smaller halos compared to Gram-positive bacteria, such as Sthaphylococcus *aureus* and *Sreptococcus pyogenes*.

The variations in determining the antimicrobial activity of plant extracts can be linked to many factors. These include the method used. According to Pereira et al. (2010), these techniques and their variants should be adapted and chosen according to the operational facilities that provide relatively quick and simple tests, at a lower cost. It is important that the adaptations are tested and standardized in order to achieve reproducibility and thus ensure that the results are reliable.

Medeiros, Fonseca and Adreata (2004) consider plants to be "the identity of a group of people, reflecting who they are, what they think and their relationship with the nature that surrounds them. This wise nature offers them food, medicine, profitable sustenance and the enjoyment of the soul". Even though we live in a mainly technological age, information about plants in folk medicine needs to arouse more and more scientific interest for use in the preparation of medicines.

References

ABU-SHANAB. et al. Antibacterial activities of some Plant extracts utilized in popular medicine in palestine. **Turkish Journal Biology**, v. 28, n. 99, p. 102, 2004.

ALVES, E.S.; TRESMONDI, F.; LONGUI, E. L. Structural analysis of *Eugenia uniflora* L. (Myrtaceae) leaves collected in rural and urban environments, SP, Brazil. **Acta Botanica Brasilica**, v. 22, n. 1, p. 241-248, 2008.

ALVES, T. M. A. et al. Biological Screening of Brazilian Medicinal Plants. **Memórias do Instituto Oswaldo Cruz,** v. 95, n. 3, p.367-373, 2000.

AURICCHIO, M.T.; BACCHI, E.M. Leaves *of Eugenia uniflora* L. (pitanga: pharmacobotanical, chemical and pharmacological properties. **Revista Instituto Adolfo Lutz**, v. 62, n. 1, p. 55-61, 2003.

AURICCHIO, M.T. et al. Antimicrobial and antioxidant activities and toxicity of *Eugenia uniflora.* **Latin American Journal of Pharmacy**, v. 26, n. 1, p. 76, 2007.

BETONI, J. E. C et al. Synergism between plant extract and antimicrobial drugs used on *Staphylococcus aureus* diseases. **Memórias do Instituto Oswaldo Cruz**, v. 101, n. 4, *p*. 387-90, jun. 2006.

BEZERRA, N.A. **Avaliaçâo da atividade antimibrobiana de Eugenia uniflora L.** Trabalho de Conclusao de Curso; (Graduation in Biological Sciences) - Universidade Estadual da Paraiba; 2013.

BOUZADA, M. L. M. et al. Search for new antimicrobial drugs from plants. **Principia - Caminhos da Iniciaçâo Cientifica**, v.11, 2007. Available at:

<http://www.ufjf.br/principia/edicoes-on-line/principia-caminhos-da-iniciacaocientifica-vol-9-2007>. Accessed on: 05 Nov 2015.

CECHINEL FILHO, V.; YUNES, R. A. Strategies for obtaining pharmacologically active compounds from medicinal plants: concepts on structural modification to optimize activity. **Quimica Nova**, v. 21, n. 1, p. 99-105, 1998.

CHATTOPADHYAY, D.; SINHA, B. K.; VAID, L. K. Antibacterial activity of Suzugium species. **Phytotherapy**, v.69, n.4, p.356-367, 1988.

CIMANGA, K. et al. Correlation between chemical composition and antibacterial activity of essential oils of some aromatic medicinal plants growing in the Democratic Republic of Congo. **Journal of Ethnopharmacology**, Limerick, v.79, n.2, p.213-220, 2002.

COSTA, J. G. M. et al. Chemico-biological study of the essential oils of *Hyptis martiusii, Lippia sidoides* and *Syzigium aromaticum* against *Aedes aegypti* larvae. **Revista Brasileira de**

Farmacognosia, v. 15, n. 4, p. 304-309, 2005.

COWAN, M. N. Plant products as antimicrobial agents. **Clinical Microbiology Reviews** v. 12, p.564 582, 1999.

DIDRY, N.; DUBREUIL, L.; TROTIN, F. and PINKAS, M. Antimicrobial activity of aerial parts of Drosera peltata Smith on oral bacteria. **Journal of Ethnopharmacology**, v.60, n. 2, p.91-96, 1998.

GONÇALVES, A.L; ALVES, A.F.; MENEZES, H. Comparative study of the antimicrobial activity of extracts from some native trees. **Arquivos do Instituto Biológico**, v. 72, n. 3, p. 353-358, 2005.

GONCALVES, F. A et al. Antibacterial activity of guava, *Psidium guajava Linnaeus*, leaf extracts on diarrhea-causing enteric bacteria isolated from Seabob Shrimp. **Revista do Instituto de Medicina** Tropical, v. 50, n. 1, p. 11-15, 2008.

HOLETZ, F. B. et al. Screening of some plants used in the Brazilian folk medicine for the treatment of infectious diseases. **Memórias do Instituto Oswaldo Cruz**, v. 97, n.7, p. 1027-1031, 2002.

KHAN, M.R. et al. Antimicrobial activity of *Symplocos cochinchinensis*. **Fitoterapia**, v.72, n.7, p.825-828, 2001.

LIMA, M.L.S. de et al. ANTIMICROBIAN ACTIVITY OF JAMBOLO (*Syzygium cumini*), THROUGH THE AGAR DIFFUSION TECHNIQUE. **Infarma-Pharmaceutical Sciences**, v. 23, n. 7/8, p. 3-7, 2012.

LOCHER, C.P. *et al.* Anti-microbial activity and anti-complement activity of extracts obtained from selected Hawaiian medicinal plants. **Journal of ethnopharmacology**, v. 49, n. 1, p. 23-32, 1995.

LÓPEZ, P. V. A. **Bioprospecting of extracts from *Croton urucurana* Baillon and its endophytic fungi**, 138f. Dissertation (Master's Degree in Microbiology) - Federal University of Paranà, UFPR, Curitiba. 2010.

LOPES, D.C.D.X.P. et al. Antimicrobial and phototoxic activities of fruit and root extracts of *Physalis angulata* L. **Revista brassileira de farmacognosia**, v. 16, n. 2, p. 206-210, 2006.

LOUGUERCIO, A. P et al. Antibacterial activity of hydro-alcoholic extract of jambolao leaves (*Syzygium cumini* (L.) Skells). **Ciencia Rural**, v. 35, n.2, p.373, 2005.

MENEZES, M. C.; et al. in vitro evaluation of the antimicrobial activity of Brazilian plant extracts on bacteria isolated from the oral cavity of dogs. **Revista Uniiversidade Rural**, v. 24, n. 2, p. 141-144, 2004.

MEDEIROS, M.F.T.; FONSECA, V.S. da; ANDREATA, R.H.P. Plantas medicinais e seus usos pelos sitiantes da Reserva Rio das Pedras, Mangaratiba, RJ, Brasil. **Acta Botanica Brasilica**, v. 18,

n. 2, p. 391-399, 2004.

MELO, R.R. et al. **Phytochemical profile, evaluation of antimicrobial activity and biocompatibility of *Syzygium malaccense* (L) Merr. & LM Perry (Myrtaceae)**. Dissertation (Master's Degree in Pharmaceutical Sciences) - Federal University of Pernambuco 2009.

OLIVEIRA, C.B.; SOARES, D.G.S.; PADILHA, W.W.N. **In vivo study on the effectiveness of decontamination of toothbrushes by the use of Eugenia uniflora L. (Pitanga) essential oil spray.** Brasilian Oral Research (Proceedings of the [24th] Annual Meeting of the Brazilian Society of Dental Research), 2007.

PEREIRA, A. V et al. Comparative study of *Momordica charantia Linn.* and *Psidium guajava Linn.* extracts on bovine *Staphylococcus aureus* strains isolated in the state of Paraiba. **Revista Agropecuària Tècnica,** *v. 31, n. 2, p. 20-28, 2010.*

PESSINI, G.L.; HOLETEZ, F.B., SANCHES, N.R.; CORTEZ, D.A.G; DIAS-FILHO, B.P.; NAKAMURA, C.V. Evaluation of the antibacterial and antifungal activity of plant extracts used in folk medicine. **Revista Brasileira de Farmacognosia** n. 13, p. 21-24, 2003. MOREIRA, V.C; FREIRE, D. *Klebsiella pneumoniae* and its resistance to antibiotics.In: MOSTRA DE PRODUÇÂO CIENTiFICA DA PÔS-GRADUAÇÃO LATU-SENSU DA PUC DE GOIAS, 6. 2011, Goiàs. **Electronic proceedings...** Goiàs:PUC, 2011. Available at:<http://www.cpgls.ucg.br/6mostra/Artigos4.html>. Accessed on: November 21, 2015.

SANCHES, N.R. et al. An evaluation of antibacterial activities of *Psidium guajava* (L.). **Brazilian Archives of biology and Technology**, v. 48, n. 3, p. 429-436, 2005.

SCHAPOVAL, E. E. S. et al. Evaluation of some pharmacological activities on *Eugenia uniflora* L. **Journal of Ethnopharmacology**. v. 44, p. 137-142, 1994.

SILVA, N. C. C. **Comparative study of the antimicrobial action of extracts and essential oils from medicinal plants and synergism with antimicrobial drugs.**2010. 75f. Dissertation (Master's Degree in General and Applied Biology) - Universidade Estadual Paulista. Botucatu, Sao Paulo, 2010.

SHAFI, P.M.; ROSAMMA, M.K.; JAMIL, K.; REDDY, P.S. Antibacterial activity of *Syzygium cumini* and *Syzygium travancoricum* leaf essential oils. **Phytotherapy**, v. 73, p. 414-416, 2002.

TURATTI, K.F.M. ***Syzygium cumini* (L.) Skeels: quality assessment, morphological and anatomical study, antimicrobial, preservative, genotoxic and mutagenic activity, cytotoxicity and incorporation into cosmetic formulations for topical use**. 184 f. Thesis (doctorate) - Universidade Estadual Paulista, Faculdade de Ciências Farmacêuticas, 2008.

VALE, V. V.; ORLANDA, J. F. F. Antimicrobial activity of the ethanolic crude extract of the aerial parts of *Euphorbia tirucalli* Linneau (Euphorbiaceae). **Scientia Plena**, v. 7, n. 4, 2011.

CHAPTER 5

RESULTS ON THE ANACARDIACEACAE FAMILY

In the analysis of the aqueous extract data, *Schinus terebinthifolius* showed antimicrobial activity against *Pseudomonas aeruginosa* and *Staphylococcus aureus* with a halo of 8 mm and 9 mm respectively at a concentration of 200 mg/ml each, but no antimicrobial activity was found in any of the concentrations of the aqueous extract of any of the other plants used. Table 1 summarizes the results of the antibiosis tests with the aqueous extract.

Table 1 - Antimicrobial activity of aqueous extracts of the leaves of *Anacardium occidentali*, *Mangifera indica, Spondias purpurea* and *Schinus terebinthifolius* at different concentrations.

Extracts	**Concentrations of aqueous extract (mg.mL)**			**CN**	**CP**
Pseudomonas aeruginosa					
	Halo of inhibition (mm)				
	200mg/ml	100mg/ml	50mg/ml		
Anacardium occidentale	0 mm	0 mm	0 mm	0 mm	15 mm
Mangifera indica	0 mm	0 mm	0 mm	0 mm	16 mm
Spondias purpurea	0 mm	0 mm	0 mm	0 mm	15 mm
Schinus terebinthifolius	8 mm	7 mm	0 mm	0 mm	16 mm
Staphylococcus aureus					
Anacardium occidentale	0 mm	0 mm	0 mm	0 mm	21 mm
Mangifera indica	0 mm	0 mm	0 mm	0 mm	21 mm
Spondias purpurea	0 mm	0 mm	0 mm	0 mm	21 mm
Schinus terebinthifolius	9 mm	8 mm	7 mm	0 mm	21 mm

- Positive control (PC): Meropenem.

- Negative control (NC): 0.9% sodium chloride.

For Lima et al. (2007), the aqueous extract of *S. terebenthifolius* expressed its activity against *Staphylococcus aureus*, S. *epidermidis*, *Bacillus cereus* and *Pseudomoas aeruginosa,* producing inhibition halos, on average, 11mm in diameter. This agrees with the results of this study, which indicated the antimicrobial activity of aroeira in aqueous extract against *Staphylococcus aureus and Pseudomoas aeruginosa.*

Aqueous extracts of siriguela leaves are reported to have antibacterial properties (AGRA et al., 2007, AYOKA et al., 2008). This differs from the results found in this study with regard to the aqueous extract.

According to Cajado (2015), the aqueous extract of *Anacardium occidentale* showed no growth inhibition against *Escherichia coli* and *Klebsiella pneumoniae*, unlike the hydroalcoholic extract, which showed better antimicrobial activity. These results confirm those obtained in this study, which showed greater susceptibility of Gram-positive bacteria when submitted to *Anacardium occidentale* leaf extract.

Cajado (2014) when analyzing the aqueous extract of *Mangifera* indica did not observe any potential for any of the concentrations 150, 75 and 37.5 mg, while when evaluating the hydroalcoholic extract, a halo of inhibition was observed on *Staphylococcus aureus* for all the concentrations studied, ranging from 9 to 9.5 mm. This differs from the hydroalcoholic extract, which in this study proved to be negative, confirming the absence of activity in the aqueous extract.

Figure 5 shows the formation of halos on *Staphylococcus* aureus using aroeira in its aqueous concentration, showing a halo of 9 mm at a concentration of 200 mg/ml.

Figura 1 - Antimicrobial activity using the Agar diffusion well technique against the *Staphylococcus aureus* species when subjected to the hydric extracts of *Schinus terebinthifolius* leaves at different concentrations.

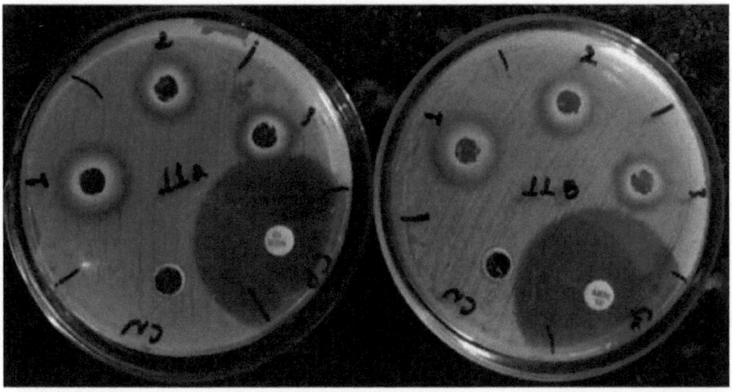

In the analysis of the data from the hydroalcoholic extracts, *Schinus terebinthifolius* showed antimicrobial activity with a halo of 8 mm at a concentration of 200 mg/ml on *Pseudomonas aeruginosa*, but no antimicrobial activity was found in any of the concentrations of the hydroalcoholic extract of any of the other plants used. On *Staphylococcus aureus*, *Anacardium occidentale* and *Schinus terebinthifolius* showed activity with halos of 9 mm and 11 mm respectively at a concentration of 200 mg/ml each. Table 2 summarizes the results of the antibiosis tests with the hydroalcoholic extract.

Table 2 - Antimicrobial activity of hydroalcoholic extracts of the leaves of *Anacardium occidentali, Mangifera indica, Spondias purpurea* and *Schinus terebinthifolius* at different concentrations.

Extracts	**Hydroalcoholic extract concentrations> (mg.mL)**			**CN**	**CP**
	Pseudomonas aeruginosa				
	Halo of inhibition (mm)				
	200mg/ml	100mg/ml	50mg/m	1	
Anacardium occidentale	0 mm	0 mm	0 mm	0 mm	14 mm
Mangifera indica	0 mm	0 mm	0 mm	0 mm	15 mm

Spondias purpurea	0 mm	0 mm	0 mm	0 mm	15 mm
Schinus terebinthifolius	8 mm	7 mm	0 mm	0 mm	16 mm
Staphylococcus aureus					
Anacardium occidentale	9 mm	8 mm	7 mm	0 mm	21 mm
Mangifera indica	0 mm	0 mm	0 mm	0 mm	22 mm
Spondias purpurea	0 mm	0 mm	0 mm	0 mm	21 mm
Schinus terebinthifolius	11 mm	10 mm	9 mm	0 mm	21 mm

- Positive control (PC): Meropenem.

- Negative control (NC): 0.9% sodium chloride.

For Silva et al. (2007), the hydroalcoholic extract of the cashew tree (*Anacardium occidentale* Linn.) produced significant antimicrobial activity *in vitro* on *Staphylococcus aureus* strains of human hospital origin. Thus, we can understand that the cashew tree presents itself as an effective therapeutic alternative for infections caused by *Staphylococcus aureus*, and is low cost and easily accessible to the population, since its use in folk medicine is already widespread.

Greatti (2014) showed that the hydroalcoholic extract of the leaves of aroeira and the essential oil of the fruit of the tree showed antibacterial activity on E. coli, in agreement with Santos (2011), using the micro dilution method, the dilutions of extracts of the leaves and stem bark of *Anacardium occidentale* showed antibacterial activity, and that Gram-positive bacteria proved to be more sensitive to the concentrations of the two parts of the plant by the diffusion technique. These results confirm those obtained in this study, which showed greater susceptibility of Gram-positive bacteria when subjected to the plant leaf extract.

Garcia and Orlanda (2014) studied the crude hydroalcoholic extract of *Mangifera indica* and found it to have antimicrobial activity *in vitro* against the *S. aureus* strain. However, it did not show antimicrobial activity against Gram-negative bacteria (*E. coli*, *P. aeruginosa* and *S. typhi*). The divergence of results for some species may be explained by the methodology used to obtain the extracts, which suggests that they have different biosynthetic and chemical classes.

For Gonçalves (2007), the hydroalcoholic extracts of seriguela leaves did not have antimicrobial activity against any of the microorganisms studied, confirming the result of the hydroalcoholic extract analyzed in this study, however, Alencar (2014) demonstrated that although the intrinsic antimicrobial activity of the *Spondias purpurea* extracts tested in his study was not considered effective according to the comparative literature, when associated with erythromycin, these extracts modified the action of the antibiotic, reducing the minimum inhibitory concentration on strains *Staphylococcus aureus* considerably.

Figure 6 shows the formation of halos on *Staphylococcus aureus* using cashew in its hydroalcoholic concentration, which showed activity with a halo of 9 mm at a concentration of 200mg/ml.

Figura 2 - Antimicrobial activity using the Agar diffusion well technique against the *Staphylococcus aureus* species when subjected to hydroalcoholic extracts of *Anacardium occidentale* leaves at different concentrations.

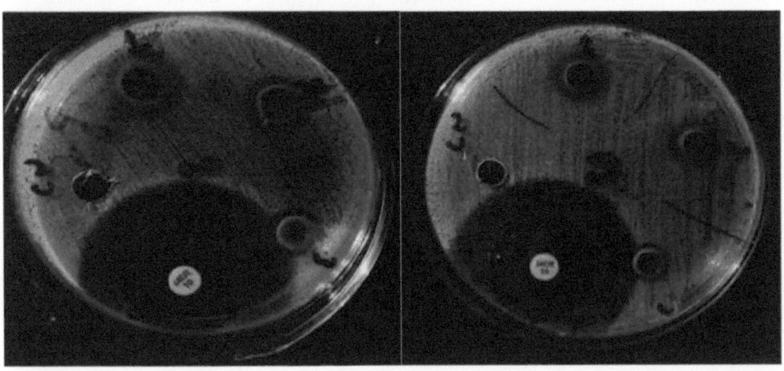

When analyzing the data from the alcoholic extracts, *Schinus terebinthifolius* showed antimicrobial activity with a halo of 9 mm at a concentration of 200 mg/ml on *Pseudomonas aeruginosa*, but no antimicrobial activity was found in any of the concentrations of the alcoholic extracts of any of the other plants used. On *Staphylococcus aureus, Anacardium occidentale, Mangifera Indica*, *Spondias*

purpurea and *Schinus terebinthifolius* showed activity with halos of 10 mm, 9 mm, 8 mm and 12 mm respectively at a concentration of 200 mg/ml each. Table 3 summarizes the results of the antibiosis tests with the aqueous extract.

Table 3 - Antimicrobial activity of alcoholic extracts of the leaves of *Anacardium occidentali, Mangifera indica, Spondias purpurea* and *Schinus terebinthifolius* at different concentrations.

Extracts	**Concentrations of alcoholic extract (mg.mL)**			**CN**	**CP**
Pseudomonas aeruginosa					
	Halo of inhibition (mm)				
	200mg/ml	100mg/ml	50mg/ml		
Anacardium occidentale	0 mm	0 mm	0 mm	0 mm	16 mm
Mangifera indica	0 mm	0 mm	0 mm	0 mm	16 mm
Spondias purpurea	0 mm	0 mm	0 mm	0 mm	15 mm
Schinus terebinthifolius	9 mm	8 mm	7 mm	0 mm	16 mm
Staphylococcus aureus					
Anacardium occidentale	10 mm	9 mm	8 mm	0 mm	21 mm
Mangifera indica	9 mm	8 mm	7 mm	0 mm	21 mm
Spondias purpurea	8 mm	7 mm	0 mm	0 mm	21 mm
Schinus terebinthifolius	11 mm	10 mm	9 mm	0 mm	22 mm

- Positive control (PC): Meropenem.

- Negative control (NC): 0.9% sodium chloride.

According to Lasca (2012), the alcoholic cashew extract showed antimicrobial activity for the disk diffusion method and was considered to have the best action, as its concentration in the evaporator route was better obtained, while the cashew extract showed antimicrobial activity for the disk diffusion and broth test methods, These results are similar to those obtained in this study, which showed different and/or more expressive inhibition halos in tests with alcoholic extracts when compared to aqueous extracts, both for cashew and for the other plants studied.

OLIVEIRA et al. (2011) investigated the synergistic activity of norfloxacin, tetracycline and erythromycin with ethanolic extract of Mangifera indica L. bark against isolates of S. aureus. The individual extract did not show significant antibacterial activity, but with the interaction with antibiotics there was a fourfold reduction in the minimum inhibitory concentration values for tetracycline and erythromycin, indicating that the extract can serve as a potential source of antibiotic adjuvant. Like the results obtained in this study, in which *Mangifera indica* showed antimicrobial activity, although not significant.

As for *Spondias purpurea*, Brito (2010) detected the a-humulene β-caryophyllene, which has an important antimicrobial action, confirming this finding only for the alcoholic extract of seriguela, but a result that is not very representative, confirming what Alencar (2014) said, showing that despite the extracts of S. purpurea having a slight antimicrobial activity, this study showed that Spondias *purpurea* had a slight antimicrobial activity. *purpurea* and *S. mombin* had antimicrobial activity, it was not considered relevant, and this study showed that *Spondias purpurea* had a slight antimicrobial activity.

Pizana et al. (2010), described that the methanolic extract of seriguela inhibited the micellar growth of the fungus *Fusarium oxysporum* by more than 50%, these data are different probably due to the type of cell wall produced by the fungus, such as cellulose or chitin, demonstrating that studies on *Spondias purpurea* can be better explored using other methodologies or even for other microorganisms.

According to Martinez, Gonzàlez and Betancourt-Badell (1996), the ethanolic extract of *S. teribinthifolius* leaves produced inhibition halos of 11 to 16 mm on agar plates against *Staphylococcus aureus*, *Bacillus subtilis*, *Pseudomoas aeruginosa* and Escherichia coli. These results may also take into account the studies by Costa (2011), where he showed that ethanolic extracts of *S. teribinthifolius* and *M. urundeuva* have antimicrobial activity against the pathogenic microorganisms *E. faecalis, Candida albicans* and *Candida tropicalis* resistant to fluconazole, also demonstrating action on fungi.

According to Freires (2010), tinctures of *S. terebinthifolius* bark (10%) were evaluated in vitro and showed antibacterial and anti-adherent activity against the microorganisms *S. mutans* and *L. casei.*

Degàspari et al. (2005) analyzed the antimicrobial potential of aqueous and alcoholic extracts obtained from the fruit of *Schinus terebinthifolius*, directly linked to the amount of phenolic compounds present in these extracts. The tests showed that the alcoholic extract had an inhibitory effect on the growth of *Staphylococcus aureus*, while the aqueous extract had no inhibitory effect on the growth of the microorganisms tested, and *P. aeruginosa* was not inhibited by any of the extracts, in disagreement with the present study, which showed activity in all the extracts and concentrations tested.

For Santos and Alves (2012), the extracts of *Anacardium occidentale* (cashew), *Schinus terebinthifolius* (red mastic) both had an inhibition halo of 24 mm. The results obtained indicate the important clinical significance of evaluating alternative and economically viable means of controlling infections caused by S. aureus.

According to Melo et al. (2006), bacterial strains of S. *mitis*, S. *mutans* and S. *sanguis* were susceptible to the action of the A. *occidentale* stem extract. Growth inhibition was homogeneous according to the degree of concentration of the plant extract. There was a proportional decrease in the diameter of the halos as the concentration of the extract was reduced, confirming the results presented in this study, which show a constant decrease in the size of the halo of inhibition as the concentration decreased.

References

AGRA, M.F. et al. Medicinal and poisonous diversity of the flora of "Cariri Paraibano". **Journal of Ethnopharmacology**, v.111, p. 383-395, 2007.

ALENCAR, E. C. B. **Avaliaçâo da atividade antimicrobiana de plantas medicinais do gènero Spondias**. (Trabalho de Conclusao de Curso apresentado ao Curso de Graduaçao em Farmâcia). p. 21. Universidade Estadual da Paraiba, 2014.

AYOKA A. O. et al. Sedative, antiepileptic and antipsychotic effects of Spondias mombin L. (Anacardiacaea) in mice and rats. **Journal of Ethnopharmacology**. v. 103, n. 2, p.166-175, 2006.

BRITO, H. R. **Chemical characterization of essential oils from Spondias mombin, Spondias purpurea and Spondias sp (cajarana do sertâo)**. (Master's dissertation in Forestry Sciences). Federal University of Campina Grande, 2010.

CAJADO, A. G. ***In Vitro* Antimicrobial Effect of Aqueous and Hydroalcoholic Extracts of Leaves of *Anacardium occidentale* and *Mangifera indica***. Specialization in Biochemistry and Applied Molecular Biology, Vale do Acaraù State University, 2014.

COSTA, C. O. D. **Avaliaçâo Da Atividade Antioxidante E Antimicrobiana De Extratos De Myracrodruon Urundeuva Allemâo E Schinus Terebinthifolius Raddi,** (Dissertaçao de Mestrado de Pós Graduaçao) Universidade Federal da Bahia, 2011.

DEGASPARI, C. H.; WASZCZYNSKYJ, N.; PRADO, µ. R. µ. Antimicrobial activity of Schinus terebenthifolius Raddi Antimicrobial activity of Schinus terebenthifolius Raddi. **Ciências e Agrotec**. Lavras. v. 29, n. 3, p. 617-622, 2005.

FREIRES, I. A. In vitro antibacterial and anti-adherent activities of tinctures of Schinus terebinthinfolius (Aroeira) and Solidago microglossa (Arnica) against dental biofilm-forming bacteria. **Odontol. Clin.-Cient**. v. 9, n. 2, 139-143, 2010.

GARCIA, A. P. M; ORLANDA, J.F.F. Evaluation of the *in vitro* antimicrobial activity of the hydroalcoholic crude extract of *Mangifera indica Linneau*. **Revista Cubana de Plantas Medicinais**. v. 19, n. 3, 2014.

GREATTI, V. R. et al. Evaluation of the in vitro antibacterial activity of mastic (*Schinus terebinthifolius*) and *cinnamon* (*Cinnamomum* zeylanicum) against Gram-positive and Gram-negative strains. **Salusvita**. v. 33, n. 3, p. 345-354, 2014.

LASCA, T. B. GONÇALVES G. M. S. **Evaluation of the Antimicrobial Effectiveness of Formulations Containing Anacardium Occidentale Extract**. Proceedings of the XVII Meeting of Scientific Initiation, Pontifical Catholic University, 2012.

LIMA. E. O. et al. *Schinus terebenthifolius* Raddi: evaluation of the antimicrobial action spectrum of its aqueous extract; **Infarma**. v.16, n. 7-8, 2004.

MARTiNEZ, M. J. A.; GONZALEZ, N.; BETANCOURT-BADELL, J. Actividad antimicrobiana del Schinus terebenthifolius Raddi (copal). **Revista Cubana de Plantas Medicinales**, v. 1, n. 3, p. 37-39, 1996.

MELO, A.F. M. de. et al. *In vitro* antimicrobial activity of *Anacardium occidentale* L. extract on

Streptococcus species. **Revista Brasileira de Farmacognosia**. v.16, n.2, p. 202-205, 2006.

OLIVEIRA, S.M.S. et al. Modulation of drug resistance in Staphylococcus aureus by extract of mango (Mangifera indica) peel. **Brazilian Journal of Pharmacognosy**, v. 21, p. 190-193, 2011.

PIZANA, C.G.; NECHA, L.L.B.; GOMEZ, Y.R. Evaluation of the fungicidal activity of leaves powders and extracts of fifteen Mexican plants against Fusarium oxysporum f.sp. gladioli (Massey) Snyder and hansen. **Plant Pathology Journal**. v. 9, n. 3, p. 103-111, 2010.

SANTOS, F. O. **Biological activities of *Anacardium occidentale* (Linn)**.(Postgraduate course in agrosilvo-pastoral systems in the semi-arid region), Federal University of Campina Grande, 2011.

SANTOS, S. J. D.; ALVES, F. Comparative analysis of the action of plant extracts with antimicrobial activity (in vitro) on strains of *Staphylococcus aureus*. **Centro Universitàrio Metodista Izabela Hendrix**, v. 02, n. 04, 2012.

SILVA, J. G. et al. Antimicrobial activity of *Anacardium occidentale* Linn. extract on multiresistant *Staphylococcus aureus* samples. **Revista Brasileira de Farmacognosia**. v.17, n.4, p. 572-577, 2007.

FINAL CONSIDERATIONS

According to the results obtained in this study for the Laminaceae family, it can be seen that the alcoholic extract showed better antimicrobial activity than the aqueous and hydroalcoholic extracts, as at least one plant species proved to be active against three types of bacteria, *S. pyogenes, S. aureus and P. aeruginosa*.

The *P. barbatus* species stands out, where its hydroalcoholic extract was active in three concentrations, 100mg/mL, 50 mg/mL and 25 mg/mL, for *S', aureus* its alcoholic extract showed antimicrobial activity against three species of bacteria, two gram-positive (*S, pyogenes, S, aureus*) and one gram-negative *(P, aeruginosa*).

It is proposed that the plant species analyzed (from the Laminaceae family) have the potential to produce constituents with antimicrobial activity against strains of *S. aureus*, thus concluding that plant products are a therapeutic alternative for the treatment of infections caused by multi-resistant bacteria and for the discovery of new drugs. Further studies are needed to verify the toxicity of the plants, isolate the active compounds responsible for the biological activity and thus expand the analysis carried out here.

The present study with species from the Anacardiaceae family showed that the alcoholic extracts of all the plants had better antimicrobial activity when compared to the aqueous and hydroalcoholic extracts. Using the Agar diffusion well technique, the largest inhibition halos occurred with the alcoholic extracts of *Anacardium occidentale* and *Schinus terebinthifolius* against the *Staphylococcus aureus* species, The largest inhibition zone for the hydroalcoholic extract tests was 10 mm for *Anacardium occidentale* at a concentration of 200 mg/ml and 12 mm for *Schinus terebinthifolius* at a concentration of 200 mg/ml, although *Mangifera indica* and *Spondias purpurea* also showed antimicrobial activity with the alcoholic extracts, but to a lesser extent.

These results suggest that the plant products analyzed represent a potential for synthesizing molecules with antibacterial activity in infectious processes caused by *Staphylococcus aureus* and *Pseudomoas aeruginosa*, but more in-depth studies to confirm this hypothesis are needed .

The present study with the Myrtaceae family supports the following conclusions: The present study showed that the alcoholic extracts of the species under study had better antimicrobial activity and a broader spectrum of action when compared to the aqueous and hydroalcoholic extracts.

Using the Agar diffusion well technique, the largest halos of inhibition occurred with the hydroalcoholic and aqueous extracts of *Eugenia uniflora* against the *Staphylococcus aureus* species.

All the plant extracts studied showed activity against *Staphylococcus aureus*, except for the alcoholic extract of *Syzygium malaccense.*

The alcoholic extract of *Psidium guajava* and *Syzygium cumini* showed activity against all the strains used in the research.

However, this study does not allow us to state that these plant species can be used as an alternative against the other bacterial species evaluated. Therefore, it is suggested that further studies be carried out with solvents of different polarities, involving the phytochemical profile, biocompatibility, chemical and biological aspects and the development of herbal medicines.

A better knowledge of the antimicrobial activity of these plants would certainly provide a boost, both directly and indirectly, for improving the living conditions of populations, considering the importance of medicinal plants, not only as a therapeutic resource, but also as a source of economic resources.

MIX
Papier aus verantwortungsvollen Quellen
Paper from responsible sources
FSC® C105338

Printed by Books on Demand GmbH, Norderstedt / Germany